Other Titles from the Society of Environmental Toxicology and Chemistry (SETAC):

Working Environment in Life-Cycle Assessment
Poulsen and Jensen, editors
2005

Life-Cycle Assessment of Metals
Dubreuil, editor
2005

Life-Cycle Management
Hunkeler, Saur, Rebitzer, Finkbeiner, Schmidt, Jensen, Stranddorf, Christiansen
2004

Scenarios in Life-Cycle Assessment
Rebitzer and Ekvall, editors
2004

Life-Cycle Assessment and SETAC: 1991–1999
(15 LCA publications on CD-ROM)
2003

Code of Life-Cycle Inventory Practice
de Beaufort-Langeveld, Bretz, van Hoof, Hischier, Jean, Tanner, Huijbregts, editors
2003

Life-Cycle Assessment in Building and Construction
Kotaji, Edwards, Shuurmans, editors
2003

Community-Level Aquatic System Studies—Interpretation Criteria (CLASSIC)
Giddings, Brock, Heger, Heimbach, Maund, Norman, Ratte, Schäfers, Streloke, editors
2002

Interconnections between Human Health and Ecological Variability
Di Giulio and Benson, editors
2002

Life-Cycle Impact Assessment: Striving towards Best Practice
Udo de Haes, Finnveden, Goedkoop, Hauschild, Hertwich, Hofstetter, Jolliet,
Klöpffer, Krewitt, Lindeijer, Müller-Wenk, Olsen, Pennington, Potting, Steen, editors
2002

Silver in the Environment: Transport, Fate, and Effects
Andren and Bober, editors
2002

Test Methods to Determine Hazards for Sparingly Soluble Metal Compounds in Soils
Fairbrother, Glazebrook, van Straalen, Tararzona, editors
2002

Avian Effects Assessment: A Framework for Contaminants Studies
Hart, Balluff, Barfknecht, Chapman, Hawkes, Joermann, Leopold, Luttik, editors
2001

SETAC

A Professional Society for Environmental Scientists and Engineers and Related Disciplines Concerned with Environmental Quality

The Society of Environmental Toxicology and Chemistry (SETAC), with offices currently in North America and Europe, is a nonprofit, professional society established to provide a forum for individuals and institutions engaged in the study of environmental problems, management and regulation of natural resources, education, research and development, and manufacturing and distribution.

Specific goals of the society are:

- Promote research, education, and training in the environmental sciences.
- Promote the systematic application of all relevant scientific disciplines to the evaluation of chemical hazards.
- Participate in the scientific interpretation of issues concerned with hazard assessment and risk analysis.
- Support the development of ecologically acceptable practices and principles.
- Provide a forum (meetings and publications) for communication among professionals in government, business, academia, and other segments of society involved in the use, protection, and management of our environment.

These goals are pursued through the conduct of numerous activities, which include:

- Hold annual meetings with study and workshop sessions, platform and poster papers, and achievement and merit awards.
- Sponsor monthly and quarterly scientific journals, a newsletter, and special technical publications.
- Provide funds for education and training through the SETAC Scholarship/Fellowship Program.
- Organize and sponsor chapters to provide a forum for the presentation of scientific data and for the interchange and study of information about local concerns.
- Provide advice and counsel to technical and nontechnical persons through a number of standing and ad hoc committees.

SETAC membership currently is composed of more than 5,000 individuals from government, academia, business, and public-interest groups with technical backgrounds in chemistry, toxicology, biology, ecology, atmospheric sciences, health sciences, earth sciences, and engineering. If you have training in these or related disciplines and are engaged in the study, use, or management of environmental resources, SETAC can fulfill your professional affiliation needs.

All members receive a newsletter highlighting environmental topics and SETAC activities, and reduced fees for the Annual Meeting and SETAC special publications. All members except Students and Senior Active Members receive monthly issues of *Environmental Toxicology and Chemistry* (*ET&C*) and *Integrated Environmental Assessment and Management* (*IEAM*), peer-reviewed journals of the Society. Student and Senior Active Members may subscribe to the journal. Members may hold office and, with the Emeritus Members, constitute the voting membership.

If you desire further information, contact the appropriate SETAC Office.

1010 North 12th Avenue
Pensacola, Florida 32501-3367 USA
T 850 469 1500   F 850 469 9778
E setac@setac.org

Avenue de la Toison d'Or 67
B-1060 Brussels, Belgium
T 32 2 772 72 81   F 32 2 770 53 83
E setac@setaceu.org

www.setac.org

Environmental Quality Through Science®

Ecosystem Responses to Mercury Contamination

Indicators of Change

Ecosystem Responses to Mercury Contamination

Indicators of Change

Based on the SETAC North America Workshop on
Mercury Monitoring and Assessment

14-17 September 2003
Pensacola, Florida, USA

Edited by

**Reed Harris • David P. Krabbenhoft
Robert Mason • Michael W. Murray
Robin Reash •Tamara Saltman**

Coordinating Editor of SETAC Books
Joseph W. Gorsuch
Gorsuch Environmental Management Services, Inc.
Webster, New York, USA

SETAC

CRC Press
Taylor & Francis Group
Boca Raton London New York

CRC Press is an imprint of the
Taylor & Francis Group, an **informa** business

CRC Press
Taylor & Francis Group
6000 Broken Sound Parkway NW, Suite 300
Boca Raton, FL 33487-2742

First issued in paperback 2019

© 2007 by Taylor & Francis Group, LLC
CRC Press is an imprint of Taylor & Francis Group, an Informa business

No claim to original U.S. Government works

ISBN-13: 978-1-58488-661-7 (SETAC Press)
ISBN-13: 978-0-8493-8892-7 (hbk)
ISBN-13: 978-0-367-38940-6 (pbk)

Library of Congress Cataloging-in-Publication Data

Ecosystem responses to mercury contamination : indicators of change / edited by Reed Harris ... [et al.].
    p. cm.
    Includes bibliographical references (p.).
    ISBN 0-8493-8892-9 (alk. paper)
    1. Mercury--Environmental aspects. 2. Environmental indicators. 3. Environmental monitoring. I. Harris, Reed.

QH545.M4E26 2006
577.27'5663--dc22
        2006049169

Visit the Taylor & Francis Web site at
http://www.taylorandfrancis.com

and the CRC Press Web site at
http://www.crcpress.com

SETAC Publications

Books published by the Society of Environmental Toxicology and Chemistry (SETAC) provide in-depth reviews and critical appraisals on scientific subjects relevant to understanding the impacts of chemicals and technology on the environment. The books explore topics reviewed and recommended by the Publications Advisory Council and approved by the SETAC North America, Latin America, or Asia/Pacific Board of Directors; the SETAC Europe Council; or the SETAC World Council for their importance, timeliness, and contribution to multidisciplinary approaches to solving environmental problems. The diversity and breadth of subjects covered in the series reflect the wide range of disciplines encompassed by environmental toxicology, environmental chemistry, and hazard and risk assessment, and life-cycle assessment. SETAC books attempt to present the reader with authoritative coverage of the literature, as well as paradigms, methodologies, and controversies; research needs; and new developments specific to the featured topics. The books are generally peer reviewed for SETAC by acknowledged experts.

SETAC publications, which include Technical Issue Papers (TIPs), workshop summaries, newsletter (*SETAC Globe*), and journals (*Environmental Toxicology and Chemistry* and *Integrated Environmental Assessment and Management*), are useful to environmental scientists in research, research management, chemical manufacturing and regulation, risk assessment, and education, as well as to students considering or preparing for careers in these areas. The publications provide information for keeping abreast of recent developments in familiar subject areas and for rapid introduction to principles and approaches in new subject areas.

SETAC recognizes and thanks the past coordinating editors of SETAC books:

Andrew Green, International Zinc Association
   Durham, North Carolina, USA
C.G. Ingersoll, Columbia Environmental Research Center
   US Geological Survey, Columbia, Missouri, USA
T.W. La Point, Institute of Applied Sciences
   University of North Texas, Denton, Texas, USA
B.T. Walton, US Environmental Protection Agency
   Research Triangle Park, North Carolina, USA
C.H. Ward, Department of Environmental Sciences and Engineering
   Rice University, Houston, Texas, USA

Contents

Chapter 3   Monitoring and Evaluating Trends in Sediment and Water
                Indicators..47

David Krabbenhoft, Daniel Engstrom, Cynthia Gilmour, Reed Harris,
James Hurley, and Robert Mason

Chapter 4   Monitoring and Evaluating Trends in Methylmercury
                Accumulation in Aquatic Biota 87

James G. Wiener, R.A. Bodaly, Steven S. Brown, Marc Lucotte, Michael C.
Newman, Donald B. Porcella, Robin J. Reash, and Edward B. Swain

Marti F. Wolfe, Thomas Atkeson, William Bowerman, Joanna Burger,
David C. Evers, Michael W. Murray, and Edward Zillioux

Preface

This book proposes a framework for a national-scale program to monitor changes in mercury concentrations in the environment following the reduction of atmospheric mercury emissions. The book is the product of efforts initiated at a workshop held in Pensacola, Florida, in September 2003, involving more than 30 experts in the fields of atmospheric mercury transport and deposition, mercury cycling in terrestrial and aquatic ecosystems, and mercury bioaccumulation in aquatic food webs and wildlife. Participants represented government agencies, industry groups, universities, and nonprofit organizations.

In many parts of North America, mercury concentrations in fish are high enough to cause concern for people and wildlife that eat fish. As a result, fish consumption advisories are common, and several states and the U.S. federal government have passed rules to reduce mercury emissions in the United States. A carefully designed monitoring program is needed to establish trends in mercury concentrations in the environment and to identify the influence of changes in mercury emissions on these trends. The charges assigned to the workshop participants included 1) the development of a set of indicators to determine whether mercury concentrations in air, land, water, and biota are changing systematically with time; 2) guidance regarding a monitoring strategy to assess these trends; and 3) guidance regarding additional monitoring needed to determine whether observed changes in mercury concentrations are related to reductions in mercury emissions. The resulting framework described in this book reflects the consensus of the workshop participants that monitoring trends in mercury concentrations at a national scale is difficult but achievable, and monitoring should be started sooner rather than later.

Acknowledgments

The authors and editors of this book wish to acknowledge the U.S. Environmental Protection Agency and the Electric Power Research Institute, who sponsored a Society of Environmental Toxicology and Chemistry (SETAC) workshop in September 2003 in Pensacola, Florida. More than 30 international experts gathered to discuss and propose a framework for a national mercury monitoring program to evaluate the effectiveness of mercury emissions controls on mercury concentrations in the environment. This book and a companion journal publication (Mason et al. 2005) are the products of the workshop and subsequent efforts.

We also wish to thank the Society of Environmental Toxicology and Chemistry (SETAC), as well as Greg Schiefer, in particular, who did an excellent job in providing the venue and organizational expertise for this project.

Each of the contributions in this book has been peer-reviewed. The opinions expressed in this book are those of the participants and may not reflect those of any of their agencies, the funding agencies, or SETAC.

About the Editors

Reed Harris is a principal engineer with Tetra Tech Inc. and has more than 25 years of experience in the environmental engineering field. Since 1988, he has focused on studying the behavior of mercury in the environment. He has developed and applied simulation models of mercury cycling and bioaccumulation in lakes, reservoirs, and the Florida Everglades. Reed is currently managing a whole ecosystem mercury addition experiment known as the Mercury Experiment to Assess Atmospheric Loadings in Canada and the United States (METAALICUS) in Ontario, Canada, that is examining the relationship between atmospheric mercury deposition and fish mercury concentrations.

David Krabbenhoft, PhD, is a research scientist with the U.S. Geological Survey. He has general research interests in the geochemistry and hydrogeology of aquatic ecosystems. Krabbenhoft began working on environmental mercury cycling, transformations, and fluxes in aquatic ecosystems with the Mercury in Temperate Lakes project in 1988; since then, the topic has consumed his professional life. In 1994, he established the USGS Mercury Research Laboratory, which includes a team of multidisciplinary mercury investigators. The laboratory is a state-of-the-art analytical facility strictly

dedicated to the analysis of mercury, with low-level speciation. In 1995, he initiated the multi-agency Aquatic Cycling of Mercury in the Everglades (ACME) project. More recently, Dave has been a Primary Investigator on the internationally conducted METAALICUS project, which is a novel effort to examine the ecosystem-level response to loading an entire watershed with mercury. The Wisconsin Mercury Research Team is currently active on projects from Alaska to Florida, and from California to New England. Since 1990, he has authored or co-authored more than 50 papers on mercury in the environment. In 2006, Krabbenhoft served as the co-host for the 8th International Conference on Mercury as a Global Pollutant in Madison, Wisconsin.

Robert P. Mason, PhD, is a professor in the Department of Marine Sciences at the University of Connecticut. Prior to this recent appointment (from September 2005), he was at the Chesapeake Biological Laboratory, part of the University of Maryland's Center of Environmental Science, for 11 years. Prior to this, he received his PhD from the University of Connecticut in 1991, and completed a postdoctoral program at the Ralph Parsons Laboratory at MIT. He has been working on various aspects of mercury biogeochemical cycling and bioaccumulation for the past 15 years and has published more than 70 papers, including numerous book chapters, on mercury in the ocean, atmosphere, and in terrestrial ecosystems. He has graduated 11 MS and PhD students during his career. His work has been widely cited and has been used to develop global mercury models and as the basis for setting local, regional, and national mercury regulations.

Michael W. Murray, PhD, has been staff scientist with the Great Lakes office of the National Wildlife Federation since 1997. His work has included scientific and policy research on a number of diverse issues involving toxic chemicals and water quality, including mercury sources, fate and transport, ecological and human health effects, and control options; assessments of water quality criteria and total maximum daily load plans; and assessment of fish consumption advisory development and communication protocols. Murray received MS and PhD degrees in water chemistry from the University of Wisconsin–Madison, where his research addressed several aspects of the environmental chemistry of polychlorinated biphenyls. He has authored or co-authored 6 peer-reviewed publications as well as numerous reports, and has served on a number of conference planning and technical committees, including the SETAC North America Technical Committee. He is also an adjunct lecturer in Environmental Health Sciences at the University of Michigan's School of Public Health.

Robin J. (Rob) Reash is a principal environmental scientist for American Electric Power, Water & Ecological Resource Services Section, in Columbus, Ohio. His principal duties include designing and conducting technical studies for NPDES compliance issues, evaluating the development of water quality standards at the federal and state levels, and conducting applied research. He has extensive experience in evaluating the effects of power plant discharges on environmental receptors (thermal effects, trace metal speciation and effects, bioaccumulation of mercury and selenium). Reash has previous work experience with the Oklahoma Water Resources Board and Ohio USEPA. He is a member of the Society of Environmental Toxicology and Chemistry and currently serves as a board member for the Ohio Valley Chapter of SETAC. He serves on 2 project subcommittees for the Water Environment Research Foundation. He has served as a peer reviewer for the USEPA proposed water quality criteria and currently serves on a panel of USEPA's Science Advisory Board. Reash received his MS degree from the Ohio State University. He has authored or co-authored 23 technical papers, and has authored 3 book chapters. In 1998, Reash was certified as a Certified Fisheries Scientist by the American Fisheries Society.

Tamara Saltman has an MS degree in marine studies (biology and biochemistry) from the University of Delaware and a BS degree in natural resource management from Cornell University. She has been helping to bridge the science and policy of environmental mercury contamination for 7 years, including facilitating the development of mercury deposition monitoring sites and communicating the results of scientific knowledge on the movement of mercury through terrestrial and aquatic environments. She also has experience developing and running a volunteer water monitoring

network, setting up a tribal water quality analysis laboratory, and training new volunteer monitors. She is currently an environmental policy analyst with the U.S. Environmental Protection Agency.

1 Introduction

Reed Harris, David Krabbenhoft, Robert Mason,
Michael W. Murray, Robin Reash, and
Tamara Saltman

How will mercury concentrations in air, land, water, and biota respond to changes in mercury emissions? This book proposes a framework for a carefully designed national-scale monitoring program, necessary but currently not in place in the United States, to help answer this question.

Mercury concentrations in many regions of the globe have increased as a result of industrial activities. Mercury contamination can occur as a localized issue near points of release and as a longer-range transboundary issue arising from atmospheric emissions, transport, and deposition. Most of the mercury (Hg) released to the environment is inorganic, but a small fraction is converted by bacteria to methylmercury (MeHg), a toxic organic compound. This is important because methylmercury bioaccumulates through aquatic food webs so effectively that most of the mercury in fish is methylmercury and fish consumption is the primary exposure pathway for methylmercury in humans and many wildlife species.

While methylmercury occurs naturally in the environment, it is reasonable to expect that methylmercury levels have increased in modern times as a result of increased inorganic mercury concentrations. Whether methylmercury concentrations have increased to a similar extent as inorganic mercury is not known. It is clear, however, that elevated fish mercury concentrations can currently be found in remote lakes, rivers, reservoirs, estuaries, and marine conditions, typically in predators such as sportfish at the top of food webs. As of 2003, 45 states had fish consumption advisories related to mercury, and 76% of all fish consumption advisories in the United States were at least partly related to mercury (USEPA 2004a). The number of advisories is increasing with time, although this is due at least partly to more sites being sampled (Wiener et al. 2003).

Regulations controlling mercury releases have been proposed or put in place for major sectors of the U.S. economy releasing mercury to the environment, including a recent rule to control emissions of mercury from coal-fired boilers (the Clean Air Mercury Rule [CAMR], USEPA 2005). Many scientists and policy makers are concerned, however, that existing monitoring programs do not provide an adequate baseline of mercury concentrations in the environment to compare against future trends or evaluate the effectiveness of emissions controls. There is significant natural variability in time and space for mercury concentrations in many environmental media, caused by a range of factors affecting mercury cycling and accumulation in biota (Figure 1.1). Local watershed and site conditions can exert large influences on

1

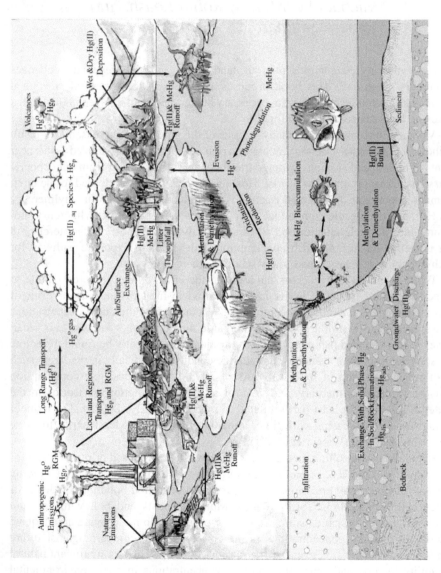

FIGURE 1.1 Conceptual diagram of mercury cycling and bioaccumulation in the environment.

mercury concentrations, as can year-to-year, seasonal, and even daily variations in meteorology. Large-scale environmental changes such as acid deposition, land use, or climate change also have the potential to enhance methylmercury production and contribute to higher fish mercury concentrations. Occasional sampling of mercury levels at a few locations is not adequate to distinguish the benefits of emissions controls from other confounding factors. There are existing long-term networks in North America that monitor wet mercury deposition, including the Mercury Deposition Network (MDN), but MDN was not designed specifically to evaluate the effects of emissions controls. For example, the majority of locations chosen for MDN sites are intentionally removed from local sources, do not provide a complete view of anthropogenically related deposition, and do not monitor dry mercury deposition rates, an important component of overall atmospheric mercury deposition.

The overall result is the current absence of a national mercury monitoring network needed to evaluate the effectiveness of regulatory actions on mercury levels in the environment and subsequent risks to humans and wildlife. In response, the U.S. Environmental Protection Agency and the Electric Power Research Institute (EPRI) sponsored a Society of Environmental Toxicology and Chemistry (SETAC) workshop in Pensacola, Florida, in September 2003 to convene more than 30 experts on this issue from North America and Europe. The purpose of the workshop was to begin the process of designing a national mercury monitoring strategy, designed to help evaluate the effectiveness of mercury emissions controls on mercury concentrations in the environment. This book and a companion journal publication by Mason et al. (2005) are the products of the workshop and subsequent efforts.

1.1 MERCURY EMISSIONS AND DEPOSITION

Anthropogenic mercury emissions to the atmosphere originate from a variety of sources, including coal combustion, waste incineration, chlor-alkali facilities, and other industrial and mining processes. Mercury emissions from these sources are not typically monitored directly. Instead, indirect methods are used, such as combining emission factors with rates of production of goods or consumption of materials, or using extrapolation methods that scale up from a limited number of sampling stations to broader national or global fluxes. Anthropogenic and naturally emitted mercury can be deposited and re-emitted repeatedly, complicating efforts to distinguish mercury emitted naturally from anthropogenically mobilized mercury. Recent estimates of natural mercury emissions, direct anthropogenic emissions, and re-emitted anthropogenic emissions suggest that these 3 "sources" are comparable (see review by Seigneur et al. 2004), totaling on the order of 6000 to 6600 metric tons per year. If these estimates are correct, mercury of anthropogenic origin would currently contribute roughly two thirds of annual mercury emissions to the atmosphere, either directly or via re-emission (Figure 1.2).

Slemr et al. (2003) attempted to reconstruct the global trend of atmospheric mercury concentrations from direct measurements since the late 1970s, and suggested that atmospheric mercury concentrations increased in the late 1970s to a peak in the 1980s, then decreased until the mid-1990s, and have been nearly constant since then. The authors noted, however, that this trend is not consistent with

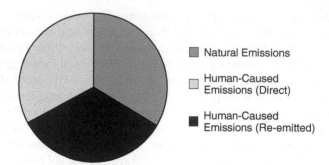

FIGURE 1.2 Estimated contributions of natural and human-caused emissions to global mercury emissions. (*Source:* From USEPA 2004b.)

inventories of anthropogenic emissions that suggest substantial global emissions reductions in the 1980s. They concluded that there is a need to improve the mercury emission inventories and to re-evaluate the contribution of natural sources.

While there is uncertainty regarding overall global trends for atmospheric mercury emissions, it is clear that the worldwide distribution of mercury emissions has been changing as some countries industrialize or invoke measures to reduce mercury releases (Figure 1.3, Figure 1.4, and Figure 1.5). North American and European anthropogenic mercury emissions declined between 1990 and 1995 (Pacyna et al. 2003), while emissions were increasing in other regions (e.g., Asia). As of 1995, approximately 10% of the total anthropogenic global mercury emissions originated in North America, while slightly more than half originated in Asia (Pacyna et al. 2003).

Other evidence also indicates that atmospheric mercury deposition rates have increased in modern times. In many remote watersheds in North America, the rate of mercury accumulation in lake sediments has increased by a factor of 2 to 5 since the mid-1800s, based on analyses of dated cores of sediment and peat (Swain et al. 1992; Lockhart et al. 1995; Lucotte et al. 1995; Lorey and Driscoll 1999; Lamborg et al. 2002). Some cores also show evidence of recent declines in mercury deposition, possibly associated with decreasing regional emissions of anthropogenic mercury (Engstrom and Swain 1997; Benoit et al. 1998). A similar picture emerged from ice cores in the Upper Fremont Glacier in Wyoming (Schuster et al. 2002), where anthropogenic mercury accounted for 70% of the accumulation in the past 100 years, although accumulation rates have been declining since the mid-1980s. Some locations are very likely more influenced by local or regional mercury sources than others. Therefore, some sites in the United States could currently be experiencing declines in mercury deposition while others are increasing.

1.2 MERCURY CONCENTRATION TRENDS IN FISH

Fish are often the focal point of interest for methylmercury contamination, representing the main exposure pathway for humans and wildlife. Unfortunately, long-term data sets with records of both mercury deposition and fish mercury concentrations over time are limited. In Sweden, Johansson et al. (2001) estimated that

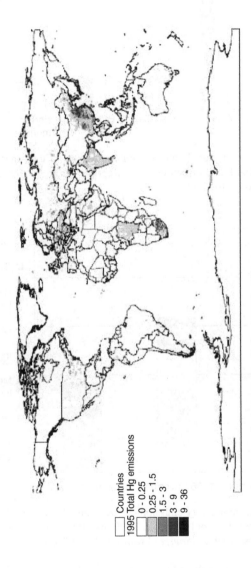

FIGURE 1.3 Anthropogenic emissions of total mercury in 1995 (tonnes). (Reprinted with permission from Pacyna et al. 2003.)

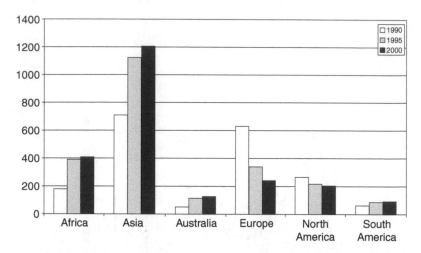

FIGURE 1.4 Change of global anthropogenic emissions of total mercury to the atmosphere from 1990–2000 (metric tons). (Reprinted from Pacyna et al. 2006, with permission from Elsevier.)

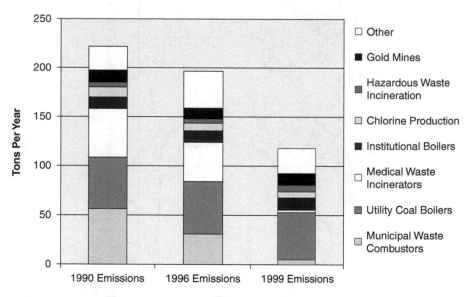

FIGURE 1.5 Anthropogenic mercury emissions in the United States, 1990–1999. Short tons per year. Emissions shown for gold mines in 1990 and 1996 are assumed to be equal to emissions for those mines in 1999. (*Source:* From USEPA 2004c.)

mercury concentrations in standardized 1-kg pike declined by 20% on average between the periods 1981–1987 and 1991–1995, possibly in association with reduced emissions from continental Europe. In North America, Hrabik and Watras (2002) concluded that fish mercury concentrations in Little Rock Lake in northern Wisconsin decreased by roughly 30% between 1994 and 2000 due to decreased atmospheric mercury loading. De-acidification was also suggested to account for an additional 5 to 30% reduction (the lake has 2 basins, 1 of which was experimentally acidified). At some sites in the Florida Everglades (e.g., site WCA 3A-15), mercury concentrations in largemouth bass have declined since the mid-1990s, perhaps as much as 60% (Atkeson et al. 2003). The Florida case is particularly relevant to several themes presented in this book. Even in an area where observations of wet mercury deposition rates began earlier than in most regions of the country, the record may be missing an important period circa 1990, when mercury deposition rates may have been higher (Pollman et al. in preparation). This illustrates the need to start monitoring networks sooner rather than later. Furthermore, the Everglades constitute a very dynamic system with many factors changing simultaneously. Sulfate concentrations in surface waters at some sites have dropped dramatically in recent years. Separating the effects of mercury deposition and sulfate loading trends is not simple, as both potentially affect methylmercury production and levels in fish (see Chapter 3 of this volume). Carefully designed monitoring programs are needed to distinguish the effects of various factors simultaneously affecting fish and wildlife mercury concentrations.

1.3 BOOK OBJECTIVES

This book is designed with 3 primary objectives:

1) Establish a set of indicators that could be monitored in the United States, and preferably North America, to determine whether mercury concentrations in air, land, water, and biota are changing systematically with time.
2) Provide guidance regarding a monitoring strategy to achieve the above goal.
3) Provide guidance regarding additional monitoring needed to help determine whether observed changes in mercury concentrations are related to regulatory controls on mercury emissions.

Geographically, this book focuses on the continental United States, although a monitoring program with a North American scope would have advantages. Emphasis is also given to systems expected to be more sensitive to changes in mercury deposition, and to freshwater and estuarine/coastal environments rather than the open oceans. It should also be noted that this book seeks to provide practical guidance, but is not a finalized detailed sampling program with specific locations, dates, frequencies, and costs.

There are 4 core chapters, distinguished by the environmental compartments on which they focus:

Chapter 2: Air/watersheds
Chapter 3: Water/sediments
Chapter 4: Aquatic biota
Chapter 5: Wildlife

Each chapter recommends indicators to monitor as a measure of changing mercury concentrations in the environment, and describes the process used by the authors to identify and rank these indicators. The chapters also discuss monitoring strategies and ancillary data needed to help interpret the extent to which atmospheric mercury deposition influences mercury concentration trends.

A final chapter (Chapter 6) provides an integrated perspective for a national mercury monitoring program, based on information from the 4 core chapters. It also recognizes that costs are a critical consideration, and offers 2 different types of assessment programs. One program focuses on documenting changes or trends in mercury concentrations in the environment, while the second is expanded in scope to also examine the impact of atmospheric mercury deposition rates on any observed changes in concentrations.

Several common themes emerged during the original Pensacola workshop. These include:

- Challenges establishing baseline mercury concentrations and temporal trends
- Challenges isolating the influence of changes in atmospheric emissions on mercury concentrations in the environment
- The benefits of a sampling strategy involving several regions nationally, each with 2 types of monitoring sites: a) intensive studies for a small set of sites, at least one in each region monitored; and b) less intensive sampling at a larger number of clustered sites in each region
- The need for coordinated monitoring studies spanning several environmental compartments through time and space, and the need for common sampling and analytical protocols; this is particularly important when striving to establish links between mercury emissions and methylmercury levels in biota
- Making use of existing datasets and coordinate with ongoing monitoring programs where possible
- The need to integrate monitoring with model development and testing.

The core chapters of the book treat these issues comprehensively, but they are introduced here briefly:

1.3.1 ESTABLISHING BASELINE CONDITIONS AND TEMPORAL TRENDS

Temporal and spatial variability impose demands on sampling programs when establishing baseline concentrations at a given site, or across a range of sites. Existing monitoring programs have shown that observations from one site cannot be considered representative nationally, nor even regionally. Even when monitoring a single site, mercury concentrations in some environmental media can vary widely between years or over short time periods, for example, in rivers where particulate mercury loads can increase dramatically during storm events. Similarly, atmospheric mercury concentrations in the vicinity of point sources may change dramatically depending on the wind direction. Mercury concentrations can also vary spatially within some environmental compartments at a given site and time. This can occur, for example, in sediments sampled only meters apart, due to heterogeneity among samples, or for a set of individual fish sampled on a given date (same species, similarly sized) due to differences in the characteristics and behavior of individual fish that generate natural variability. The authors concisely explore these obstacles in this book, and offer strategies to address them.

1.3.2 ESTABLISHING CAUSE-EFFECT RELATIONSHIPS

It is not a simple matter to show a cause-effect relationship between mercury emissions and methylmercury concentrations in biota. Individual sites are impacted by different mixes of near-field, mid-range, and long-range mercury emissions sources. The chapter authors also discuss confounding factors beyond mercury loading that can influence total and methylmercury concentrations in the environment, including atmospheric, terrestrial, and aquatic chemistry; land use and urbanization; hydrology; climate change; and trophic conditions. Ecosystems also exhibit a range of ecosystem sensitivities and response dynamics to changes in mercury deposition. Some systems may respond faster than others or have variable rates of response (e.g., relatively quickly at first but slower later). As a result, different temporal trends may emerge at different locations, thus complicating efforts to isolate the effects of mercury emissions and deposition on fish mercury concentrations.

These considerations require an expanded scope for a monitoring program, involving measurements of ancillary environmental conditions in addition to mercury data if the objective is not just to document changes in mercury concentrations, but also to gain insight into links between emissions and concentration trends in biota. These issues are addressed in each chapter.

1.3.3 SAMPLING STRATEGY

Two basic sampling strategies available are to 1) carry out limited sampling at many sites, or 2) carry out intensive sampling at a smaller set of sites. Both strategies provide benefits, although they differ. Focusing resources on a small number of sites provides a more accurate picture of what is happening at those few locations, and is well-suited to developing a better mechanistic understanding of processes and links between mercury deposition and mercury concentrations in biota (e.g., the Mercury in Temperate Lakes (MTL) programs in the late 1980s and early 1990s (Watras et al. 1994). Distributing resources across a wide range of sites can provide a more regional or national perspective, but at the price of confidence in what is being actually being observed at any one location. As a result of these trade-offs, the authors present a combined strategy involving clusters of sites, some sampled more intensively, distributed across different regions nationwide.

1.3.4 MONITORING DATA AND MODELING

Policy makers would benefit from a combination of strong field evidence of trends and well-established models to draw upon when assessing the benefits of past or future policy decisions. Models of mercury cycling and bioaccumulation are not yet adequately predictive across a range of conditions and landscapes. Results from a national mercury monitoring program, if carefully designed, offer the potential to help develop models of mercury cycling and bioaccumulation. The more intensively sampled sites in particular could prove useful to advance the capability of models. Opportunities to link monitoring with models are discussed in the core chapters.

Overall, the behavior of mercury is too complex to easily establish the benefits of emissions controls. A carefully designed monitoring program is needed, involving not just mercury data, but also a suite of carefully selected environmental parameters spanning several environmental compartments. The remainder of this book provides guidance toward reaching this difficult but achievable goal.

REFERENCES

Atkeson T, Axelrad D, Pollman C, Keeler G. 2003. Recent trends in mercury emissions, deposition and concentrations in biota. In: Integrating Atmospheric Mercury Deposition and Aquatic Cycling in the Florida Everglades: An Approach for Conducting a Total Maximum Daily Load Analysis for an Atmospherically Derived Pollutant. Integrated Summary Final Report. Florida Department of Environmental Protection (FDEP), Tallahassee, FL. http://www.floridadep.org/labs/mercury/index.htm

Benoit JM, Fitzgerald WF, Damman AWH. 1998. The biogeochemistry of an ombrotrophic bog: evaluation of use as an archive of atmospheric mercury deposition. Environ Res (Sect A) 78:118–133.

Engstrom DR, Swain EB. 1997. Recent declines in atmospheric mercury deposition in the upper Midwest. Environ Sci Technol 31(4):960–967.

Hrabik TR, Watras CJ. 2002. Recent declines in mercury concentration in a freshwater fishery: isolating the effects of de-acidification and decreased atmospheric mercury deposition in Little Rock Lake. Sci Total Environ 297:229–237.

Johansson K, Bergbäck B, Tyler G. 2001. Impact of atmospheric long range transport of lead, mercury and cadmium on the Swedish forest environment. Water, Air Soil Pollut: Focus 1:279–297.

Lamborg CH, Fitzgerald WF, Damman AWH, Benoit JM, Balcom PH, Engstrom DR. 2002. Modern and historic atmospheric mercury fluxes in both hemispheres: global and regional mercury cycling implications. Global Biogeochem Cycles 16(4):1104.

Lockhart WL, Wilkinson P, Billeck BN, Hunt RV, Wagemann R, Brunskill GJ. 1995. Current and historical inputs of mercury to high-latitude lakes in Canada and to Hudson Bay. Water, Air Soil Pollut 80(1–4):603–610.

Lorey P, Driscoll CT. 1999. Historical trends of mercury deposition in Adirondack lakes. Environ Sci Technol 33:718–722.

Lucotte M, Mucci A, Hillaire-Marcel C, Pichet P, Grondin A. 1995. Anthropogenic mercury enrichment in remote lakes of northern Québec (Canada). Water Air Soil Pollut 80:467–476.

Mason RP, Abbott ML, Bodaly RA, Bullock Jr OR, Driscoll CT, Evers D, Lindberg SE, Murray M, Swain EB. 2005. Monitoring the response of changing mercury deposition. Environ Sci Technol 39:14A–22A.

Pacyna EG, Pacyna JM, Steenhuisen F, Wilson D. 2006. Global anthropogenic mercury emission inventory for 2000. Atmospheric Environment 40:4048–4063.

Pacyna JM, Pacyna EG, Steenhuisen F, Wilson S. 2003. Mapping 1995 global anthropogenic emissions of mercury. Atmos Environ 37(Suppl. 1):S109–S117.

Pollman CD, Porcella DB, Engstrom DR. (In preparation). Assessment of trends in mercury-related data sets and critical assessment of cause and effect for trends in mercury concentrations in Florida biota: phase II.

Schuster PF, Krabbenhoft DP, Naftz DL, Cecil LD, Olson ML, Dewild JF, Susong DD, Green JR, Abbott ML. 2002. Atmospheric mercury deposition during the last 270 years: a glacial ice core record of natural and anthropogenic sources. Env Sci Technol 36:2303–2310.

Seigneur C, Vijayaraghavan K, Lohman K, Karamchandan P, Scott C. 2004. Global source attribution for mercury deposition in the United States. Environ Sci Technol 38:555–569.

Slemr F, Brunke EG, Ebinghaus R, Temme C, Munthe J, Wangberg I, Schroeder W, Steffen A, Berg T. 2003. Worldwide trend of atmospheric mercury since 1977. Geophys Res Lett 30(10):1516.

Swain EB, Engstrom DR, Brigham ME, Henning TA, Brezonik PL. 1992. Increasing rates of atmospheric mercury deposition in midcontinental North America. Science, New Series 257:784–787.

[USEPA] US Environmental Protection Agency. 2005. Standards of performance for new and existing stationary sources: electric utility steam generating units; Final Rule. Fed Reg 70, Wednesday, May 18, 2005/Rules and Regulations. 40 CFR Parts 60, 72, and 75. [OAR-2002-0056; FRL-7888-1]. RIN 2060–AJ65

[USEPA] US Environmental Protection Agency. 2004a. Fact Sheet — National Listing of Fish Advisories. Office of Water EPA-823-F-04-016 August 2004. URL: http://www. epa.gov/waterscience/fish/advisories/ factsheet.pdf

[USEPA] US Environmental Protection Agency. 2004b. URL: http://www.epa.gov/mercury/ control_emissions/global.htm, updated May 2005.

[USEPA] US Environmental Protection Agency. 2004c. URL: http://www.epa.gov/mercury/ control_emissions/emissions.htm, updated December 2004.

Watras CJ, Bloom NS, Hudson RJM, Gherini SA, Munson R, Klaas SA, Morrison KA, Hurley J, Wiener JG, Fitzgerald WF, Mason R, Vandal G, Powell D, Rada R, Rislove L, Winfrey M, Elder J, Krabbenhoft D, Andren AW, Babiarz C, Porcella DB, Huckabee HW. 1994. Sources and fates of mercury and methylmercury in remote temperate lakes. In: Watras CJ, Huckabee JW, editors, Mercury pollution — integration and synthesis. Boca Raton (FL): Lewis Publishers. p. 153–177.

Wiener JG, Krabbenhoft DP, Heinz GH, Scheuhammer AM. 2003. Ecotoxicology of mercury. In: Hoffman DJ, Rattner BA, Burton Jr GA, Cairns Jr J, editors, Handbook of ecotoxicology, 2nd ed. Boca Raton (FL): CRC Press. p 409–463.

2 Airsheds and Watersheds

Charles T. Driscoll, Michael Abbott, Russell Bullock, John Jansen, Dennis Leonard, Steven Lindberg, John Munthe, Nicola Pirrone, and Mark Nilles

ABSTRACT

As a result of controls that have been recently implemented and that are proposed for atmospheric emissions of mercury (Hg), there is a critical need to design and implement a program to monitor ecosystem response to these changes. The objective of this chapter is to review the state of Hg monitoring activities and programs that are currently being conducted for airsheds and watersheds, and to make recommendations to strengthen and add to these programs in order to quantify future changes that may occur as a result of changes in atmospheric emissions of Hg and subsequent deposition. In this regard we identified a series of airshed and watershed indicators that, when measured over a long period of time, should help to determine the (response from) changes in the global, continental, and/or regional-scale Hg emissions (or other watershed loads of Hg such as land-use changes or discharges). Note that an important benefit of improved Hg monitoring programs would be the availability of high quality data to test and validate models. These data would help support the development and application of models as research tools to better understand the dynamics and cycling of Hg in complex environments. Improved and well-validated models could subsequently be used as management tools to predict the response of airsheds and watershed ecosystems to changes that might occur in emissions of Hg or other changes that might alter the transport or bioavailability of Hg (e.g., changes in atmospheric deposition, climate change, land disturbance). To achieve this objective we propose an integrated airshed/watershed Hg monitoring program. We propose that within an ecoregion detailed sampling at intensive study sites (intensive sites) and less intensive sampling at a larger number of clustered sites (cluster sites) would be conducted. To evaluate Hg response in airsheds we propose a series of air quality Hg intensive sites. At these intensive sites detailed measurements of atmospheric Hg speciation and deposition would be made together with supporting measurements of atmospheric chemistry and meteorology. Several air quality Hg intensive sites exist and could be used as templates for this approach. We also propose measurements of total ecosystem deposition at the air quality Hg intensive sites. Researchers have suggested that throughfall plus litterfall might be used as a cost-effective surrogate for total Hg deposition to forest ecosystems. While this approach needs further research, we believe it holds considerable promise and might ultimately be implemented at cluster sites. We strongly endorse that continued use of the Mercury Deposition Network (MDN). The MDN is a North American network in which wet Hg deposition is measured using standard protocols. The MDN is the only national framework that currently exists to monitor changes in Hg deposition. The MDN needs continued support and should be expanded to improve spatial coverage. For watersheds, we recommend that an intensive watershed monitoring program be initiated to measure changes in the chemistry and flux of Hg species in streamwater over the long-term. Rather than implementing a new watershed monitoring program, we recommend that a Hg monitoring component be added to existing watershed networks (i.e., the NSF LTER program, USGS WEBB program). Existing programs have the advantage of monitoring infrastructure and expertise that is already in place and a record of ancillary measurements, which would be critical to the interpretation of ecosystems response to changes in Hg deposition. At the cluster-level, we recommend that a forest floor or surface soil monitoring program be implemented to evaluate the response of soil to changes in atmospheric Hg deposition.

2.1 INTRODUCTION

There is a critical need to establish an integrated, long-term monitoring program to quantify the inputs, transport, and fate of atmospheric mercury (Hg) deposition within watershed ecosystems, and the response of Hg indicators to changes in Hg emissions, atmospheric deposition of Hg and other materials (e.g., acidic deposition), climate events or change, and/or land disturbance or change. Central to this need is the integration of approaches and data on Hg monitoring of airsheds and watersheds. We envision that the response of airsheds and watersheds to changes in Hg emissions will be variable across time and space (Figure 2.1, Figure 2.2, and Table 2.1; Engstrom and Swain 1997; Bullock and Brehme 2002). At the local scale, air chemistry and deposition near local sources should be elevated and respond rapidly to changes in local emissions of particulate mercury (PHg) and reactive gaseous mercury (RGHg). At the regional scale, sites that are within a source area but some distance (~50 km) from sources should respond, albeit to a lesser extent, to changes in emissions of PHg and RGHg. The lifetime of RGHg is short (hours to days), and RGHg concentrations observed at remote sites are primarily related to photochemical oxidation of gaseous elemental Hg (Hg(0)), most likely by reactive halogens and oxidants. Note that the conversion of Hg(0) to RGHg is enhanced near coastal regions (Pirrone et al. 2003a). Particulate Hg at remote sites is formed from similar reaction or from preexisting suspended particulate matter that adsorbs gaseous Hg. Therefore, remote sites that are far removed from emission sources should largely reflect changes in global emissions of Hg(0).

Watersheds are sinks for atmospheric Hg deposition (Grigal 2002). However, they are highly variable in their ability to retain inputs of total Hg (THg), convert ionic Hg (Hg(II)) to bioavailable methylmercury (MeHg), and supply Hg(II) and MeHg to downstream aquatic ecosystems, ultimately influencing exposure to sensitive biota and humans.

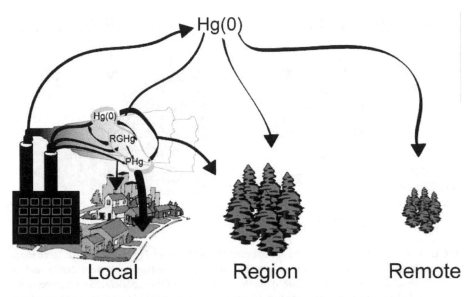

FIGURE 2.1 A conceptual diagram illustrating the sources and pathways of atmospheric Hg, and the response of deposition to changes in Hg emissions. Near sources of Hg emissions, deposition of particulate Hg (PHg) and reactive gaseous Hg (RGHg) is high and probably responsive to changes in emissions. Areas that are distant from sources but within the source area will receive lower deposition of PHg and RGHg and will be less responsive to changes in emissions. Finally, areas that are remote from sources of Hg emissions (and local and regional sites) will receive Hg deposition that largely originates from oxidation of elemental Hg (Hg(0)) from global sources. Remote sites will not be responsive to local and regional changes in emissions.

The pathways of Hg transport and sites of Hg transformations within watershed ecosystems are complex and poorly understood. Like airsheds, it is envisioned that different types of watersheds will respond differently to changes in atmospheric Hg deposition. The response of a watershed to changes in Hg deposition will be a function of hydrologic flowpaths through the watershed, climate, soils and surficial geology, vegetation type, and landscape features. For example, watersheds with urban land cover and considerable runoff from impervious surfaces should receive elevated inputs of Hg and, in the absence of confounding variables, be responsive to changes in atmospheric Hg deposition (Figure 2.2). However, urban watersheds may be influenced by land-use changes, nutrient enrichment, local point Hg sources, and other factors, which may make it difficult to discern changes solely to emission controls. Perched seepage lakes derive their waters largely from direct precipitation and shallow hydrologic flowpaths. These ecosystems should be fairly responsive to changes in atmospheric Hg deposition. In contrast, surface waters draining watersheds with thick deposits of surficial materials that strongly retain Hg might be expected to respond initially only to direct deposition to the lake surface and respond slowly or not at all to changes in atmospheric deposition of Hg to the watershed.

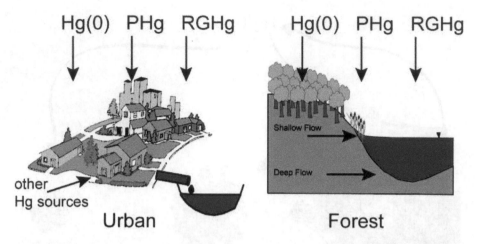

FIGURE 2.2 A conceptual diagram illustrating the response of Hg in watersheds to changes in atmospheric Hg deposition. As an example, shown is an urban ecosystem that would be responsive to deposition changes due to the short-circuiting of flow associated with impervious surfaces. Urban watersheds also are complicated by sources of Hg in addition to atmospheric deposition. A perched seepage lake would be responsive to deposition changes because water is largely derived from direct deposition to the lake surface and shallow flow paths. A lake with water derived from deep groundwater would probably not respond rapidly to changes in deposition.

TABLE 2.1
Response of 4 hypothetical lake ecosystems to changes in national Hg emissions. (Note that these concepts are also relevant for river and coastal ecosystems.)

Ecosystem type	Urban lake	Forest lake	Forest lake	Forest lake
Hg deposition	High	Moderate	Moderate	Low
Hg sources	Local, regional, global	Regional, global	Regional, global	Global
Airshed response	High, rapid	Moderate, rapid	Moderate, rapid	None
Hydrologic flowpath	Short-circuited	Shallow	Deep	N/A
Watershed response	Rapid	Moderate	Slow–none	None

Note that these concepts are also relevant to the transport of Hg to river and coastal ecosystems.

Watershed disturbance may confound the interpretation of Hg response patterns. Virtually every watershed disturbance alters the supply of THg and/or the conversion of Hg(II) to MeHg. These disturbances might include changes in atmospheric deposition, land disturbance or change, climatic events or long-term climate change, or local Hg contamination from industries or wastes. For example, clear-cutting or other land disturbances have been shown to increase watershed export of THg and

MeHg (Porvari et al. 2003; Munthe and Hultberg 2004). Also, long-term decreases in sulfate, which have occurred across Europe and eastern North America for 30 years, could alter transformations of Hg(II) and/or MeHg or the bioavailability of MeHg through changes in surface water pH, net production of dissolved organic carbon (DOC), and/or sulfate available for reduction and associated production of MeHg (Hrabik and Watras 2002). Watershed disturbances are widespread and should be addressed in the design of a watershed Hg monitoring program.

2.1.1 OBJECTIVE

The objective of this chapter is to review the state of Hg monitoring activities and programs that are currently being conducted for atmospheric Hg chemistry and deposition and watersheds in North America and Europe, and to make recommendations to strengthen these programs and establish new programs to quantify future changes that may occur due to changes in atmospheric emissions of Hg and subsequent deposition. In this regard we identified a series of airshed and watershed indicators that, when measured over a long period of time, should help determine the (response from) changes in the global, continental, and/or regional-scale Hg emissions (or other watershed loads of Hg such as land-use changes or discharges). The purview of this chapter is limited to atmospheric and watershed terrestrial indicators. Indicators associated with the aquatic, wetlands, riverine, sediment, and biotic compartments of the ecosystem are addressed in subsequent chapters of the book (see Chapters 3, 4, and 5).

Note that an important benefit of improved Hg monitoring programs would be the availability of high-quality data to test and validate models. These data would help support the development and application of models as research tools to better understand the dynamics and cycling of Hg in complex environments. Improved and well-validated models could subsequently be used as management tools to predict the response of airsheds and watershed ecosystems to changes that might occur in emissions of Hg or other changes that might alter the transport or bioavailability of Hg (e.g., changes in atmospheric deposition, climate change, land disturbance).

To achieve this objective, we propose an integrated airshed/watershed Hg monitoring program. There are 2 broad approaches that have been used previously in the design of monitoring programs. The first approach is to obtain data over a large spatial area. If sites for this spatial program are selected on a statistical basis, then it is possible to make an estimate of the population of the resource that shows a characteristic or change. This approach has been widely embraced by policymakers because it provides a quantitative framework for estimating damages or the extent of recovery following a mitigation strategy (e.g., Landers et al. 1988; Kamman et al. 2003). The disadvantage of this approach is that for a complex, highly reactive pollutant such as Hg, it is difficult to detect real changes. Moreover, without supporting data, it is difficult to determine the mechanism responsible for this change. The second approach utilizes intensive and detailed measurements at a small number of sites. With this approach it is easier to detect change and attribute this change to a mechanism, but it is difficult to know how representative this phenomenon is to the population of resources at risk. Our proposed program would utilize both approaches. Consistent with the approach discussed elsewhere (Mason et al. 2005) and this volume (Chapters 3 and 6), we propose that within an ecoregion, detailed sampling at intensive study sites (intensive sites) and less intensive sampling at a larger number of clustered sites (cluster sites) would be conducted.

2.1.2 LIMITATIONS

Because much remains to be learned about the complex relationships between emissions and deposition of Hg, between deposition and terrestrial flux of Hg to the aquatic environment, and all of the factors that affect and control such relationships, it is difficult to identify good indicators that completely meet our objective. Further-more, interpretation of changes in the indicators (i.e., trends) as to causality (i.e., from emissions changes or from changes in other controlling factors such as mete-orology) is difficult and must be performed with caution. In this regard, there are a series of limitations that must be kept in mind as one designs, implements, and interprets the results of a program to measure indicators.

2.1.2.1 Emissions of Mercury

Although atmospheric emissions (and other terrestrial loads to watersheds) of Hg were deemed outside the scope of this chapter, it is important to note that the reliability of the relationships between emissions changes and environmental indi-cators of that change can only be as good as the reliability of emission estimates. Therefore, it is recommended that quantification through research and monitoring of all Hg emissions sources (e.g., natural, anthropogenic, re-emissions) be aggres-sively pursued globally.

2.1.2.2 Detection of Trends

Mercury indicators often exhibit strong temporal and spatial variability. The ability to detect real trends in any of the recommended indicators at a single site will depend on several factors that can obscure or impart such trends to the data:

1) The consistency of methods used to measure the indicators (see Sections 2.2.3 and 2.3.3 for further discussion and recommendations regarding methods).
2) The role of meteorological and climatic factors and their variability.
3) Ambient air quality (e.g., oxidant concentrations) and deposition that can affect the emissions to indicator relationship. Sampling frequency is also an important attribute of a monitoring program that strongly influences the ability to detect trends.
4) The strength of the signal to all of the "noise" will be critical in deter-mining how readily trends can be discerned. The "strength" of the signal is generally a function of the distance from the source (Figure 2.3).

Understanding spatial variability is critical to detecting real trends across numer-ous sites. Because every site is affected differently by global, regional, and local

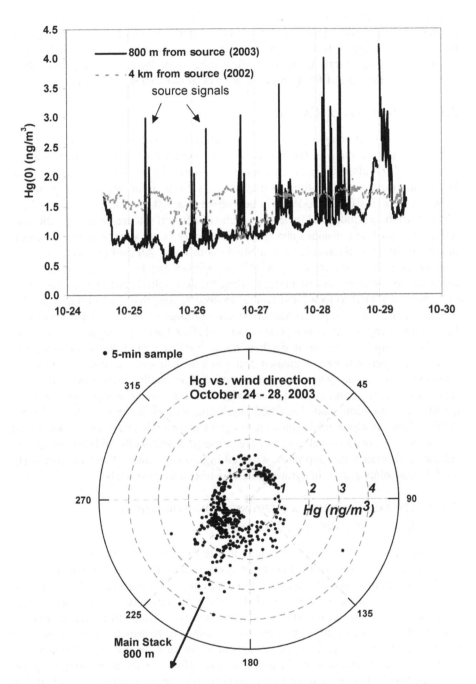

FIGURE 2.3 The ability to detect trends in atmospheric emissions can be strongly affected by the distance from the source (top) and meteorological factors such as wind direction (bottom). These measurements were made near a Hg source in southeastern Idaho. (*Source:* Abbott 2003, unpublished data, with permission.)

emissions as well as meteorological factors, the detection of trends due to any particular source will require long-term records. Critical to assessing the changes at these different scales will be locating monitoring sites in areas that are predominately impacted by atmospheric deposition originating from local, regional, and global sources (e.g., downwind from an urban area vs. remote sites).

2.1.3 ATTRIBUTION OF CAUSALITY

As in all statistical analysis, a strong correlation does not necessarily mean cause and effect. If 1 or more of the Hg indicators changes corresponding with marked changes in Hg emissions, causality cannot necessarily be assumed. If all controlling factors are measured over time, it may be possible to infer causality empirically. However, it is likely that models will be needed to assist in making the causal link. Models will be a critical tool to determine and quantify if real trends in Hg indicators are the result of an emissions change or some other factor such as meteorological- or air–quality related change, or watershed disturbance. Airshed and watershed Hg models are still in the early stages of development and testing. As a result, it is important to continue work on understanding the atmospheric chemical and physical processes, at global to local, and annual to hour scales, that control the emission-to-deposition relationship. It is also critical to continue process-level and watershed Hg studies to improve process representation and allow for the testing of watershed Hg models. To support such an understanding as well as provide the data needed to evaluate the performance of airshed and watershed models, a limited number of intensive sites measuring a comprehensive suite of air quality, meteorological water-shed variables are recommended, worldwide and in a variety of meteorological, air quality, and emissions and watershed environments (see Section 2.2.6 for further details). The most powerful approach to detect real trends in Hg indicators would be comprehensive empirical data that are consistent with well-validated model calculations. To realize this approach, we need high-quality airshed and watershed Hg models, and high-quality integrated data sets to test these models.

2.1.4 OVERALL CRITERIA FOR SELECTING MONITORING SITES, GLOBAL AND REGIONAL INFLUENCE

Historically, support for environmental monitoring networks has been sporadic. Support shifts with the political attention given to a particular environmental issue. Commonly, a phenomenon is asserted to be a major environmental problem and the lack of information that would be needed to understand its nature, extent, and impact is decried. A program of monitoring and research is instituted to gather the knowl-edge needed to develop an appropriate policy response. A response is fashioned and implemented and frequently a pledge is given to continue environmental monitoring to evaluate the effectiveness of the policy actions. However, the monitoring program associated with the issue in many cases enters into decline as new issues are identified and limited resources are demanded by other problems. In this phase, budget-driven changes (such as temporary shutdowns, site moves or closures, changes in sampling intervals, and reductions in quality assurance and quality control) diminish the value of the long-term data set due to overall loss of continuity in the historical record. The start-up and shutdown costs of designing and implementing networks are significant. The inefficiencies of such an approach add to the delays in addressing emerging issues and to the cost of generating the information required to develop sound policy.

Finally, the value of extensive time-series records extends beyond the identification of a specific problem. Long-term time-series permits verification that decisions are effective (or not); solutions are, indeed, working (or not); and the ongoing costs and benefits of the given control program are assessed accurately. With proper design of what to measure, it can also assist in understanding the why or why not.

1) *Co-location.* Extensive synergies can be gained by co-locating Hg indicator monitoring with existing networks for monitoring other important measures of air quality, deposition, and watershed characteristics. The existing networks of monitoring sites provide a low-cost infrastructure that is readily modified to include new chemical species of interest, such as Hg. The ability of emerging monitoring programs to build on an established traditional infrastructure (e.g., trained technicians, secured and well-documented sites, field laboratories) has resulted in lower start-up costs, quicker implementation schedules, and fewer initial problems for new measurement objectives. Also important for new initiatives is the ability to access the substantial knowledge-based infrastructure associated with a monitoring network, such as trained data management and quality assurance specialists, sophisticated data and site management tools, and data dissemination (e.g., interactive Internet-based servers for supplying environmental data to a worldwide customer base). Finally, the existing long-term time-series of other environmental indicators at such sites are more useful when co-located with monitoring for new constituents such as Hg indicators.

2) *Longer-term sites.* Response to long-term changes in Hg emissions can be obscured by the large day-to-day, season-to-season, and year-to-year variations in winds, temperature, precipitation hydrology, and atmospheric circulation patterns that, in turn, affect dispersion, transport, and deposition of Hg, and subsequent retention and/or transport in watersheds. To see beyond these shorter-term and random variations, it is important to select sites that have a long-term commitment and site protection to provide continuity of monitoring for long periods of time, using consistent procedures and quality assurance practices to observe long-term and significant changes in atmospheric Hg contributions to airshed and watershed response.

3) *Representative locations.* Important indicators of response to Hg emissions should be measured across a range of climatic, geographic, and watershed conditions, and encompass a range of Hg deposition regimes and not only where the greatest impacts in endpoints are expected. Continental background sites are needed to evaluate and partition global

background from natural and anthropogenic regional emissions, to initial-
ize airshed model boundary conditions, as well as to evaluate changes
primarily attributable to changes in global background levels. Sites are
also needed across a wide range of climatic, depositional, and watershed
characteristic ranges to provide data for development and performance
assessment of continental-scale models of atmospheric Hg concentrations
and deposition, and watershed-scale models of Hg fate and transport.

Sampling of Hg indicators in an urban environment are commonly distinct from
samples collected from sites deemed to be regionally representative. Urban sampling
for Hg indicators should consider the importance of defining an urban, suburban,
rural, and pristine gradient. Given the human health and wildlife endpoints for Hg,
it is important to collect information in locations representative of the environments
where fish capture and consumption is prevalent (see Chapters 4, 5). This occurrence
is common in regions considered neither remote nor strictly urban, and the response
of indicators in these regions should not be neglected. The response of indicators in
urban locations and along an urban to rural gradient is particularly important to
serve as a sensitive measure to changes in significant local emissions sources. Sites
located away from large local sources would be expected to be less responsive to
such local changes.

2.2 AIRSHEDS

2.2.1 INTRODUCTION

The concept of a watershed is easily understood. The path taken by water flowing
on the Earth's surface is determined largely by topography. However, the "airshed"
is a concept that is not so easily understood due to the 3-dimensional and time-
variant nature of atmospheric flow. The definition of an airshed is based on assumptions about wind flow patterns surrounding a location of interest and the length of
time that a substance is transported in the atmosphere. As such, airsheds cannot be
defined rigidly. This is especially true for atmospheric Hg, which exists in a number
of physicochemical forms, each of which has a different atmospheric lifetime. There
are excellent published reviews of the atmospheric chemistry and cycling of Hg to
which the reader is referred to for further details (e.g., Schroeder and Munthe 1998).
Atmospheric Hg is typically described in 3 basic forms: as Hg(0), RGHg, or
PHg. Elemental Hg is a relatively inert substance (although see Sections 2.2.2.3 and
2.2.7), minimally soluble in water, and is believed for the most part to remain in the
atmosphere for months before being deposited to the surface or chemically converted
into the more readily deposited RGHg or PHg forms. Thus, the majority of Hg(0)
emitted to air can be expected to travel globally and be mixed throughout the entire
atmosphere. RGHg and PHg are much more rapidly deposited and thus their atmospheric lifetimes are much shorter (i.e., on the order of a few days or less). Because
PHg is primarily removed by washout and RGHg is removed by both wet and dry
deposition processes, PHg has a slightly longer residence time than RGHg. As a
result, the airshed for atmospheric Hg as a whole is indeed a rather indefinite concept.

Depending on the form of atmospheric Hg, the associated airshed can vary from global to local scales.

Of the 3 atmospheric Hg species, only Hg(0) has been tentatively identified with spectroscopic methods (Edner et al. 1989), while RGHg and PHg are operationally defined (i.e., their chemical and physical structures cannot be exactly identified by experimental methods but are instead characterized by their properties and capability to be collected by different sampling equipment). Reactive gaseous Hg (RGHg) is defined as water-soluble Hg species with sufficiently high vapor pressure to exist in the gas phase. The most likely candidates for RGHg species are halogen compounds such as $HgCl_2$ and $HgBr_2$, but possibly other Hg(II) species also exist (e.g., $Hg(OH)_2$). Particulate Hg (PHg) consists of Hg bound or adsorbed to atmospheric particulate matter. Several different components are possible; Hg(0) or RGHg adsorbed to the particle surface, Hg(II) species chemically bound to the particle or integrated into the particle itself. Another species of particular interest is methyl-mercury (MeHg), due to the high capacity of this species to bioaccumulate in aquatic food chains and its subsequent role in human and wildlife exposure to Hg. MeHg is found in the atmosphere, and atmospheric deposition may substantially contribute to the MeHg loading of aquatic ecosystems (Bloom and Watras 1989; Brosset and Lord 1991; Hultberg et al. 1994; Lee et al. 2003). Because MeHg is only present at low concentrations (i.e., picogram/m^3) in ambient air, it is not an important species for the overall atmospheric cycling of Hg, but should be included because of its capacity for bioaccumulation. Gaseous methylated mercury species have recently been quantified at concentrations in landfill gas of several orders of magnitude above ambient (Lindberg et al. 2002), suggesting that direct deposition could be important near such sources. Typical concentrations of atmospheric Hg species are presented in Table 2.2.

The basic indicators for atmospheric Hg as it pertains to environmental contamination are wet and dry deposition. Concentrations of the various forms of Hg in outdoor air are rarely high enough to be of health concern via inhalation. However, air concentrations of each of the 3 basic chemical species of Hg must be obtained in order to understand their resulting behavior during transport, and rates of dry deposition to surfaces. In fact, it remains quite difficult to directly measure the dry deposition rate for Hg in any of its forms. As a substitute for direct measurement, time-integrated total (wet+dry) deposition fluxes can be determined by measuring the Hg content of throughfall and litterfall (see Section 2.2.7). However, dry deposition of Hg has been largely ignored in the deposition monitoring programs that have been initiated to date.

Variations of these indicators over time can occur due to changes in emissions, but they can also be due to meteorological variability and to changes in the measurement method employed. Signal-to-noise ratio is a critical issue for all of the indicators identified above due to the global nature of the airshed for atmospheric Hg and the regional-to-local nature of proposed emission reductions. The development and use of numerical model simulations of atmospheric Hg emissions, transport, transformations, and deposition should be used to measure pertinent indicators and to better understand the fundamental atmospheric processes that affect Hg dynamics.

TABLE 2.2
Typical concentrations of Hg species in the planetary boundary layer

Species	Concentration range	Location	Ref.
Hg(0)	0.5–1.2 ng m^{-3}	Atlantic air, southern hemisphere	
	1.1–1.8 ng m^{-3}	Atlantic air, continental background, northern hemisphere	Wängberg et al. (2001); EC (2001)
	0.8–2.2 ng m^{-3}	Mediterranean air*	Sprovieri et al. (2003); Pirrone et al. (2001, 2003a)
	1.5–15 ng m^{-3}	Continental air, urbanized, industrial	
	0.1–1.4 ng m^{-3}	Arctic*	Sprovieri, Pirrone (2000)
	0.1–1.1 ng m^{-3}	Antarctica*	Sprovieri et al. (2002); Ebinghaus et al. (2002)
	1.7–4.1 ng m^{-3}	United States	Keeler et al. (1995) Landis et al. (2002)
RGHg	<30 pg m^{-3}	Background air	
	up to 40 pg m^{-3}	Marine and continental (**)	Sprovieri et al. (2003); Pirrone et al. (2001, 2003a)
	5>50 pg m^{-3}	Near sources	Wängberg et al. (2003)
	up to 200 pg m^{-3}	Antarctica and Arctic (**)	Sprovieri et al. (2002)
PHg	–5 pg m^{-3}	Background air	
	0.1–25 pg m^{-3}	Marine (Mediterranean air) (**)	Sprovieri et al. (2003); Pirrone et al. (2001, 2003a)
	5–>50 pg m^{-3}	Continental background, higher near sources	Wängberg et al. (2003)
		Antarctica and Arctic (**)	Sprovieri et al. (2002)
CH$_3$HgX	0.1–10 pg m^{-3}	Background air	Lee et al. (2003)
(CH$_3$)$_2$Hg	<5 pg m^{-3}	Background air	
	–30 pg m^{-3} (v.v. and normally <5 pg m^{-3})	Marine polar air	
Hg(II) in precipitation	1–20 ng L^{-1}		Wängberg et al. (2001) Keeler et al. (1995)

(*) Sampling time of 5 minutes, whereas the average concentrations reported in the table are related to the whole study period.

(**) Sampling time of 2 hours, whereas the average concentrations reported in the table are related to the whole study period.

2.2.2 THE CHEMISTRY OF ATMOSPHERIC MERCURY

2.2.2.1 Dry Deposition to Terrestrial and Aquatic Receptors

Dry deposition of Hg can occur via 2 processes: 1) the direct deposition of gas-phase Hg(0), and 2) the deposition of RGHg and, to a much lesser extent, atmospheric particulate matter to which Hg is reversibly or irreversibly adsorbed. The first process is extremely difficult to quantify, depending as it does on not only meteorological phenomena such as temperature and wind speed, but also on the type and geomorphology of the surface under consideration. Nevertheless, models and several recent chamber studies indicate that vegetation has the ability to absorb Hg(0) directly from the atmosphere (Lindberg et al. 1992; Hanson et al. 1995; Frescholtz 2002). However, to simplify the system, most regional scale studies have assumed that the gaseous flux of Hg(0) over the land/water surface is zero (Pai et al. 1997; USEPA 1997; Bullock and Brehme 2002). Recently, a number of flux chamber experiments, especially on water surfaces, have been performed to test the validity of this assumption and to determine whether it is possible to parameterize net fluxes as a function of air and sea temperature and solar irradiation (Pirrone et al. 2003).

The second process, that of RGHg deposition together with particulate matter, has been addressed in various regional scale modeling studies for some time, but only recently has it been considered for direct measurement. Reactive gaseous Hg exhibits the characteristics of a so-called "sticky gas" and is commonly modeled in the same fashion as nitric acid vapor (e.g., USEPA 1997; Bullock and Brehme 2002).

These gases deposit rapidly due to their reactivity with surfaces, and exhibit elevated dry deposition velocities; rapid dry deposition has been confirmed in recent field studies in forests and the Arctic (Lindberg and Stratton 1998; Lindberg et al. 2002). At concentrations typical of rural or remote ecosystems, the dry deposition of RGHg and Hg(0) are far greater than PHg, although this species may be of importance under dry conditions near sources (Pirrone et al. 2000).

2.2.2.2 Wet Scavenging by Precipitation Events

Wet removal processes concern soluble chemical species (Hg(II)) and its compounds, and some Hg(0), and also particulate matter scavenged from within and below the precipitating clouds.

The total wet deposition flux consists of 2 contributory factors. The first derives from the continuous transfer of Hg to cloud water, described by chemistry models. There are 2 limiting factors: 1) the uptake of gas phase Hg(0), which is regulated by the Henry's constant; and 2) the subsequent oxidation of Hg(0) to Hg(II), which is governed by reaction rate constants and the initial concentrations of the oxidant species. The total flux depends on the liquid water content of the cloud and the percentage of the droplets in the cloud that reach the Earth's surface.

The second contribution to the total wet Hg flux is the physical removal of particulate matter and the scavenging of RGHg from the atmosphere during precipitation events.

2.2.2.3 Atmospheric Residence Time

Many studies have indicated an atmospheric lifetime of Hg(0) of around 1 year, based on mass balance considerations. Field measurements in the Arctic and Antarctic during polar spring have, however, shown that under these specific conditions, Hg(0) can behave as a reactive gas with a lifetime of minutes to hours (e.g., Schroeder et al. 1998; Ebinghaus et al. 2002) during Hg depletion events. These Hg depletion events occur only during a limited time of a few weeks and are not representative of the overall behavior of atmospheric Hg(0). Hedgecock and Pirrone (2004) have shown in a modeling study that atmospheric Hg(0) has the shortest lifetime when air temperatures are low, and sunlight and deliquescent aerosol particles are plentiful, which indicates that Hg(0) may have a shorter lifetime in specific circumstances other than the polar spring.

2.2.3 MEASUREMENTS AND ANALYTICAL METHODS

Sampling and analysis of atmospheric Hg is often made as total gaseous Hg (TGHg), which is an operationally defined fraction defined as species passing through a 0.45-μm filter or some other simple filtration device such as quartz wool plugs and collected on gold. Total gaseous Hg (TGHg) is mainly composed of Hg(0) vapor, with minor fractions of other volatile species such as $HgCl_2$, CH_3HgCl, or $(CH_3)_2Hg$. At remote locations, where PHg concentrations are usually low, Hg(0) is the predominant form (>99%) of the total Hg concentration in air (Table 2.2).

In the past few years, new automated and manual methods have been developed to measure TGHg (Ebinghaus et al. 1999); RGHg (Stratton et al. 2001; Feng et al. 2000; Landis et al. 2002); and PHg (Keeler et al. 1995; Lu and Schroeder 1999). These developments make it possible to determine both urban and background concentrations of RGHg, PHg, and TGHg. Accurate determinations of emissions and ambient air concentrations of different Hg species will lead to an increased understanding of the atmospheric behavior of Hg and to more precise determinations of source-receptor relationships. This information, linked with other data, can be used to assess the various pathways of human exposure to Hg (EU Commission 2001; USEPA 1997).

Denuders have been used in a variety of air pollution studies to collect reactive gases for subsequent analysis, such as ammonia, nitric acid, and sulfur oxides (Ferm 1979; Possanzini et al. 1983). Denuders were also used to remove reactive gases to prevent sampling artifacts associated with aerosol collection (Stevens et al. 1978). Gold-coated denuders were developed for removal of Hg vapor from air but were not applied to air sampling (Munthe et al. 1991). Potassium chloride (KCl)-coated tubular denuders, followed by silver-coated denuders, were used by Larjava et al. (1992) to collect $HgCl_2$ (RGHg) and Hg(0) emissions from incinerators.

For PHg, a variety of different filter methods have been applied, such as Teflon or quartz fiber filters. Before analysis, these filters undergo a wet chemical digestion usually followed by reduction-volatilization of the Hg to Hg(0) and analysis using cold vapor atomic absorbance spectrometry (CVAAS) or cold vapor atomic fluorescence spectrometry (CVAFS). Recently, a collection device based on small quartz fiber filters mounted in a quartz tube was designed. The Hg collected on the filter can be released thermally, followed by gold trap amalgamation and CVAFS detection (Lu et al. 1998; Wängberg et al. 2003).

There is a critical need to develop standard methods that can be widely adopted at national and international scales; these methods must form the basis of regional and global scale networks.

2.2.4 Modeling and the Need for Co-location/Intensive Sites

Although significant improvements in speciated measurement methods have occurred over the past decade, there are still limitations in accuracy and detection. These limitations reduce the ability to detect changes in atmospheric concentrations of Hg caused by small changes in emissions. Our understanding of critical atmospheric processes is also incomplete, and field observations can exhibit characteristics that are not readily explainable based on current scientific understanding. There remains a significant degree of uncertainty with regard to the identity and rate of various reduction/oxidation reactions of Hg in air and atmospheric water that are known to have a significant effect on its transport and deposition behavior (Arriya et al. 2002; Feng et al. 2000; Gårdfeldt and Jonsson 2003). Current numerical simulation models of atmospheric Hg are using chemical and physical reaction definitions that are almost certainly incomplete and inaccurate to some degree. Thus, great care should be taken when using models to provide source attribution for observed air concentrations and depositions of Hg or incremental changes in those parameters that would be expected to occur as future emission controls are implemented for Hg and other pollutants that might interact with it. By monitoring the concentration and deposition of other constituents at the same time and place as for Hg species, models can be further developed and tested with fewer degrees of freedom. Moreover, increased confidence can be placed on observed Hg signals that correlate with the signals for other pollutants, as expected based on current science.

2.2.5 Existing Atmospheric Mercury Monitoring Networks

In 1994, the National Atmospheric Deposition Program (NADP) Mercury Deposition Network (MDN) was established in the United States and Canada to develop a North American database on the weekly concentrations of THg in precipitation and the seasonal and annual flux of THg in wet deposition. The data are used to develop an information database on the status and spatial and seasonal trends in wet Hg deposition to surface waters, forested watersheds, and other sensitive receptors. Additional objectives are to gain a better understanding of the relation between Hg emissions and wet Hg deposition, to provide ground truth for model development, and collect baseline data to gauge the effectiveness of proposed future controls on Hg emissions. The locations of the ~85 National Atmospheric Deposition Program (NADP) Mercury Deposition Network (MDN) sites operating in 2006 in the United States, Canada, and Mexico are shown in Figure 2.4. A sub-set of approximately 20 MDN sites is sampled for MeHg concentration and deposition.

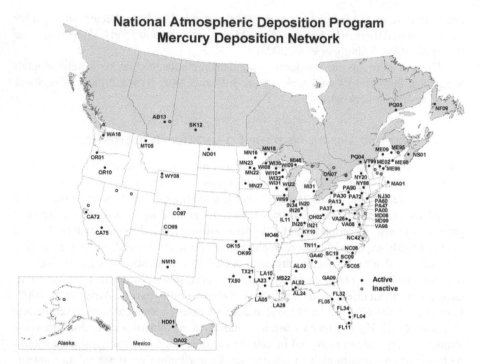

FIGURE 2.4 Location of Mercury Deposition Network (MDN) sites in 2006.

Support for sites is multi-tiered and includes participation by numerous federal, state, private, academic, and tribal organizations. Network operation includes rigorous field and laboratory quality assurance/quality control (QA/QC), including an external quality assurance program and periodic external on-site audits.

Most NADP sites meet stringent siting criteria that ensure the collection of valid precipitation samples that are regionally representative and not unduly influenced by individual local sources. A sub-set of sites is located in or near urban areas where local sources may predominate. More sites are located in regions of the United States with the most sensitive lakes and highest number of fish advisories for Hg. All sites are required to use the same sampling equipment, sampling frequency, sampling protocols, and central network laboratory. Data from all sites in the network, along with additional information on site descriptions and the overall network, can be accessed on the NADP Web site (http://nadp.sws.uiuc.edu). A summary of annual wet Hg deposition for 2004 from the MDN data is shown in Figure 2.5a. To illustrate the temporal variability in wet deposition, a time-series of THg in precipitation at 1 of the sites is shown in Figure 2.5b.

Limitations of the NADP/MDN include generally inadequate station coverage in the western United States, as well as discontinuities in areas of the eastern United States, including some areas with expected high levels of deposition and ecosystem sensitivity. NADP/MDN samples are integrated weekly precipitation samples. Unless only 1 precipitation event occurs in a given week, these integrated samples are not ideal for back-trajectory modeling and source apportionment of wet deposition.

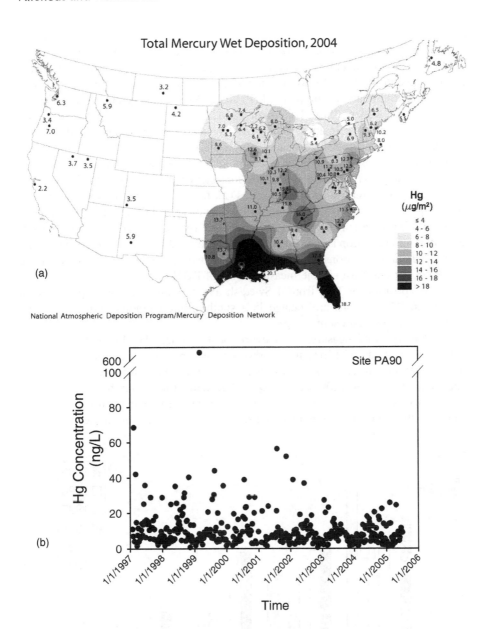

FIGURE 2.5 a) Wet THg deposition at the Mercury Deposition Network (MDN) sites for 2004 and b) temporal patterns in the concentration of THg in precipitation at a MDN site, based on weekly observations.

The considerably higher costs for operation and sample analysis of an event-based network currently preclude a higher sampling frequency. In addition, data from the networks are limited to wet deposition. Methodologies for determining dry deposition are subject to greater uncertainties, and can be more costly than wet deposition and entail significant logistical efforts to obtain. Hence, a critical need is the development of an integrated scheme for dry deposition measurement in a network mode.

In Europe, measurements of atmospheric Hg have been a voluntary part of the Co-operative Programme for Monitoring and Evaluation of the Long-Range Transmissions of Pollutants in Europe (EMEP) since 1998. Existing monitoring stations are located in Sweden, Finland, Germany, and Norway. Measurement programs are variable but generally include TGHg (automatic or manual), PHg, and wet Hg deposition. The coverage of northern Europe is satisfactory, although higher frequency measurements would be more desirable from a model validation perspective. The observed geographical patterns are consistent with the location of the main emissions source areas in central Europe. Data from the northernmost station (Finland) represent a global background with no or very little influence of European emissions. For assessment of the total deposition (wet and dry) of THg in Europe, as well as for evaluation and testing of atmospheric models of the European domain, 4 or 5 additional stations in southern Europe would be desirable. None of the European stations have routine measurements of RGHg.

Wet deposition of THg from 4 Swedish and 1 Finnish station is presented in Figure 2.6. The stations are located in a south-to-north gradient with increasing distance from the main source regions.

A globally based measurement network for atmospheric Hg does not exist. Long-term measurements of TGHg have, however, been performed at some land-based stations. Data from remote marine sites are mainly available from research cruises. Recently, Slemr et al. (2003) compiled TGHg data from a number of permanent stations and oceanic cruises for the northern and southern hemispheres (Figure 2.7). Although the data set is incomplete, a peak in TGHg concentrations seems to have

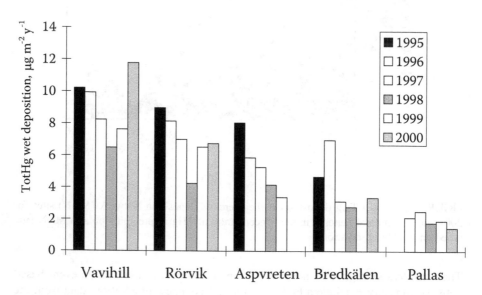

FIGURE 2.6 Wet deposition of THg over a south to north gradient in Sweden and Finland. Vavihill (southernmost) and Pallas (northernmost). Data are shown for 6 years.

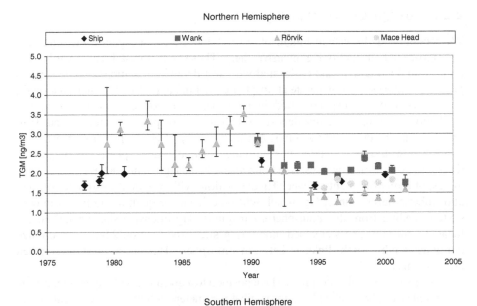

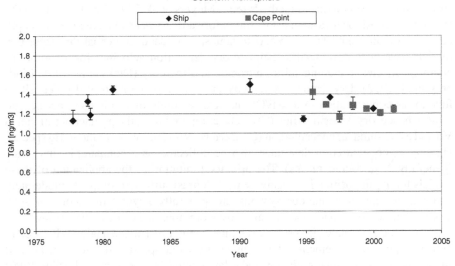

FIGURE 2.7 Time series of concentrations of total gaseous mercury (TGHg) in the northern and southern hemispheres after Slemr et al. (2003). These data were compiled from permanent monitoring stations and from oceanic cruises. Note that there may be a peak in TGHg during the 1980s. Concentrations appear higher in the northern hemisphere than in the southern hemisphere. (Reproduced with permission from the American Geophysical Union.)

occurred in the mid-1980s in the northern hemisphere. Values appear to have declined from that period to the mid-1990s. Total gaseous Hg concentrations would seem to have been constant over the past 10 years. Concentrations in the southern hemisphere are lower than in the northern hemisphere and are less complete but appear to show a similar pattern.

2.2.6 AIR QUALITY MERCURY INTENSIVE SITES

We recommend the establishment of several intensive atmospheric measurement sites (so-called air quality Hg intensive sites). These sites, located with regard to the criteria outlined above, and co-located with wet deposition stations, would serve 3 primary purposes: 1) to generate atmospheric data in support of regional-, national-, and global-scale atmospheric modeling efforts; 2) to collect data for local-scale modeling of atmospheric dry deposition to surfaces of primary interest; and 3) to improve understanding of the atmospheric chemistry of Hg. A primary objective of these sites would be collection of continuous speciated atmospheric Hg data using the standardized methods described above for PHg, Hg(0), and RGHg. These estimates, plus measurements of wet THg deposition, will provide the estimates of total atmospheric deposition of Hg, which will be critical for quantifying trends in ecosystem loading from the atmosphere. Air quality Hg intensive sites should be co-located with intensive watershed monitoring sites (see Section 2.3.2) to maximize our understanding of the linkages between atmospheric Hg deposition and watershed Hg dynamics.

Estimation of total atmospheric Hg deposition to a given site will require monitoring of a number of chemical and meteorological parameters, which are now readily measurable through standardized methods (e.g., Meyers et al. 1998; Landis et al. 2002). While wet deposition measurement is straightforward, dry deposition requires the application of existing models to appropriate atmospheric data. Inferential dry deposition models have been developed to estimate dry deposition velocities for a number of chemical species, including sulfur, nitrogen, and mercury (Hicks et al. 1991; Lindberg et al. 1992). Dry deposition velocity is defined as the ratio of dry deposition flux to air concentration ($Vd = F/C$), and carries units of centimeters per second (cm/s); hence, dry deposition flux (F) can be computed from the product of a modeled Vd and a measured air concentration (C). Because Vd varies widely among Hg species, Hg species concentration data must also be collected on a comparable time scale (generally 2- to 4-hour means). The primary limitation to the application of these models for dry deposition is a relatively well-defined surface in simple to moderately complex terrain; mountainous systems are generally beyond the scope of such models but could be addressed using ecosystem approaches described below. The models, which simulate air/surface exchange using a resistance analog (Hicks et al. 1987), require measurements of instantaneous wind speed and direction, air and surface temperature, solar radiation, and relative humidity (Meyers et al. 1989, 1996) along with measurements of atmospheric Hg speciation. The models require information on the surface of interest (e.g., grassland, forest, water, bare soil), including the plant species if present, the distribution of leaf area with height, and stomatal characteristics. The output of model calculations includes prediction of Vd and fluxes of chemical species of interest on the time scale of hours. Such a system is now in place for sulfur and nitrogen species at the AirMon sites in the United States, where weekly wet deposition measurements are combined with modeled output to routinely generate weekly estimates of total atmospheric deposition (http://www.arl.noaa.gov/research/programs/airmon.html). We feel that the level of confidence developed from using this approach over the past decade will allow reasonable estimates to be made of total Hg deposition on weekly, seasonal, and annual time scales.

Several air quality Hg intensive sites exist and could be used as templates to determine what additional air quality measurements should be included in evaluating the performance of air quality models. These include the USEPA SuperSite programs (http://www.epa.gov/ttn/amtic/supersites.html) and the Southeastern Aerosol Research and Characterization (SEARCH) project (http://www.atmospheric-research.com/studies/SEARCH/index.html).

2.2.7 TOTAL ECOSYSTEM DEPOSITION

We also recommend the establishment of a program for "direct" measurement of total Hg deposition at the ecosystem level. A number of authors have suggested that total Hg deposition to forests may be considerably higher than wet deposition (Driscoll et al. 1994; Munthe et al. 1995a, 1995b; Lindberg 1996; Rea et al. 2001; St. Louis et al. 2001), and modeled estimates of dry Hg deposition appears to be comparable to, or much larger than wet deposition (Lindberg et al. 1992; Bullock and Brehme 2002). It is clear that wet deposition is not an accurate reflection of total atmospheric loading to many surfaces, and, by itself, is probably insufficient to indicate trends. To improve our ability to estimate total Hg loading, independent ecosystem-level deposition estimates are necessary. Moreover, these measurements would provide independent validation of modeled Hg fluxes estimated at the proposed air quality Hg intensive sites. The recommended program would be most useful if co-located with the intensive-measurement catchments discussed in the next section. Each site would include a sub-set of the equipment from the air quality Hg intensive sites with which weekly wet and dry Hg deposition would be determined (as described above, but accomplished with a cheaper, less-intensive sampling approach for Hg species and meteorological variables than needed at the air quality Hg intensive sites).

Several studies of Hg fluxes in forests have indicated that the total atmospheric deposition of Hg might be estimated from its fluxes in throughfall and litterfall (Driscoll et al. 1994; Munthe et al. 1995a, 1995b; Lindberg 1996; Rea et al. 2001; St. Louis et al. 2001; Figure 2.8). Throughfall (TF) is the rain that passes through the vegetation canopy washing off accumulated dry deposition, and is an excellent indicator of seasonal total deposition of air pollutants that are relatively inert in the canopy and for which root uptake and canopy leaching are minor (e.g., Lindberg and Garten 1988). Mercury fluxes in TF generally exceed those in rain, suggesting dry deposition washoff (e.g., Driscoll et al. 1994; Lindberg 1996; Rea et al. 2001). However, at many sites, the most significant flux of Hg to the forest floor occurs in litterfall (LF), which far exceeds wet deposition (Driscoll et al. 1994; Munthe et al. 1995a, 1995b; Lindberg 1996; Rea et al. 2001; St. Louis et al. 2001; Figure 2.8). If this Hg represents an atmospheric source, and is not the result of soil uptake, LF can be used as a component of estimates of total deposition (Johnson and Lindberg 1995). Several recent studies support this interpretation (including temporal trends of Hg in foliage, controlled gas-exchange studies, and soil Hg-uptake experiments (e.g., Lindberg 1996; Rea et al. 2001; St. Louis et al. 2001; Erickson et al. 2003;

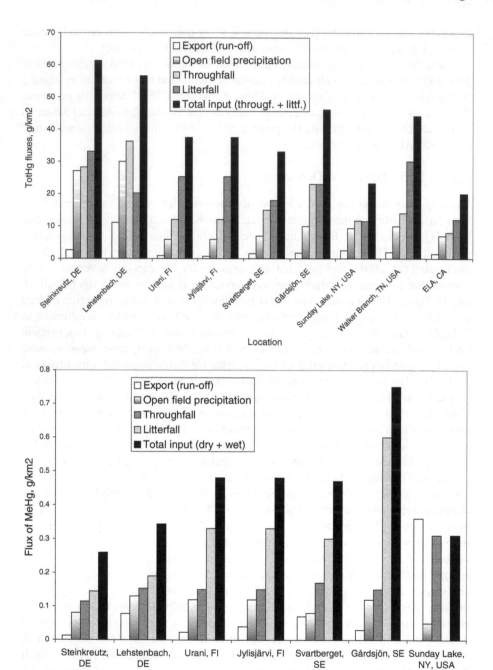

FIGURE 2.8 Inputs and losses of a) THg and b) MeHg for watersheds in Europe and North America.

Frescholtz 2002). Although ongoing and new planned field and laboratory studies are designed to further test this hypothesis, we feel that it is warranted at this time to develop a pilot-scale network of annual ecosystem fluxes of THg in TF and LF as indicators of total atmospheric deposition. These fluxes can then be compared with measured wet plus modeled dry deposition based on both inferential and regional-scale models to develop independent estimates of total atmospheric deposition for forested catchments. We also believe that this approach could eventually be applied to a national network, such as the MDN. Although this method is best aimed at forested sites, ongoing research will address methods appropriate for other ecosystems.

2.2.7.1 Snow Surveys

In western North America, persistent winter snowpack provides an excellent sampling medium to detect both spatial and temporal changes in atmospheric Hg deposition. Snowpack is an efficient integrator of both wet and dry atmospheric Hg deposition and often provides direct Hg input to streams and lakes with little soil interaction, which can complicate the link between atmospheric deposition and aquatic end-points. Widespread persistent snowpacks in North American coastal ranges and throughout the Rocky Mountains provide a sampling medium for evaluation of both trans-Pacific inputs of Eurasian pollutants to the continent (Wilkening et al. 2000) and spatial and temporal trends in local and regional source impacts (Susong et al. 2003; Abbott et al. 2002; Figure 2.9). Because Hg re-emission loss can occur over time in snowpack (Lalonde et al. 2002), care should be exercised in the evaluation of snowpack concentrations that have been sampled at different intervals after snow-fall events. This re-emission loss, however, will likely result in end-of-season snow-pack data, providing a measure of the net total depositional input to runoff, which is the primary Hg input to some lakes.

Short-term in-season temporal trends in Hg deposition can be investigated by coupling 10-cm snowpack interval concentrations with snowfall event dates from nearby SNOwpack TELemetry (SNOTEL) sites. In addition, potential source directions during deposition events can be determined by examining wind directions during the snowfall events or using back-calculated modeling trajectories. Long-term trends may be investigated by sampling the same sites over several winters.

Estimates of THg loading ($\mu g/m^2$) can be determined for a dated snowpack interval by the product of the interval Hg concentration and the snow water equiv-alent, which is determined from the interval density and thickness. Seasonal and, at many sites, near annual (~90%) loadings can then be estimated by summing the interval loadings (USEPA 2003).

2.3 WATERSHEDS

2.3.1 Introduction

Watersheds integrate the signal of atmospheric deposition and define the interface between the atmosphere and many aquatic ecosystems. The primary indicators

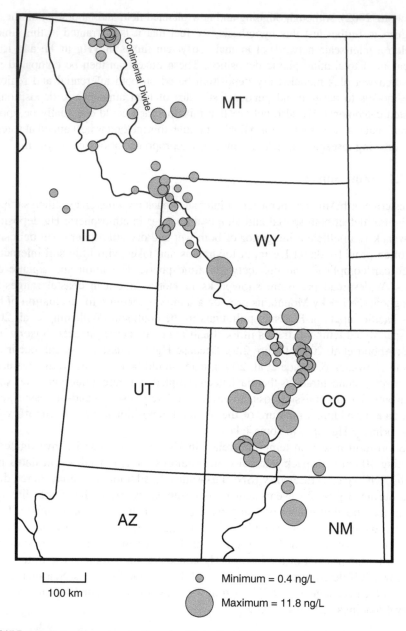

FIGURE 2.9 Total Hg concentrations (ng/L) in the 2002 snowpacks at snow-sampling sites in the Rocky Mountains of the United States (GP Ingersoll and others, U.S. Geological Survey, written commun., 2003).

reflecting the role of the entire watershed to retain atmospheric Hg deposition and supply THg and MeHg to downstream aquatic ecosystems are the concentrations and fluxes of Hg species in surface water. There are other potential indicators that may be helpful tools in assessing the spatial extent of changes in atmospheric Hg deposition. Note that concentrations or fluxes of THg and MeHg may be influenced by other factors in addition to current THg deposition.

Watershed transport and transformations of Hg are poorly studied. Inputs of Hg are lost by volatilization, soil sequestration, and drainage. Based on a review of the literature, Grigal (2002) estimated rates of volatilization of ~38 $\mu g/m^2$-yr, soil sequestration ~5 $\mu g/m^2$-yr, and stream loss of ~2 $\mu g/m^2$-yr. Although all values are highly uncertain and variable across ecosystems, soil Hg(0) volatilization is particularly poorly characterized. Stream concentrations and flux of THg appear to be weakly and inversely related to watershed size. Particulate matter and DOC are important carriers of stream Hg; any factors that influence the loss of these materials will affect Hg transport. Most studies have reported low stream fluxes of MeHg (<0.15 $\mu g/m^2$-yr). Watersheds exhibiting elevated MeHg transport are characterized by wetlands. Reducing conditions and extended hydraulic residence time make wetlands ideal sites for MeHg production and transport to interconnected lake ecosystems. Grigal (2002) provides a critical review of Hg transformations and transport in watersheds.

A few studies in which Hg mass balances were developed for watersheds suggest that there is a lack of a clear linkage between annual Hg deposition and catchment loss of either THg or MeHg. In fact, recent studies in Sweden and Finland indicate that the export of THg may be more strongly influenced by catchment disturbance than by small-scale changes in atmospheric Hg deposition (Munthe and Hultberg 2004; Porvari et al. 2003). Moreover, recent controlled Hg loading studies at the Experimental Lakes Area (ELA), Canada, suggest that the THg exported in any given year is probably derived largely from native soil pools of Hg, rather than new Hg deposition. Therefore, measurements of catchment export of Hg must be regarded cautiously with respect to the use of this measurement as a reliable indicator of trends in atmospheric deposition.

In this section, we discuss the concentrations and fluxes of THg and MeHg exported from watersheds as a result of atmospheric deposition and other sources. We also recommend a monitoring program to be implemented at a series of intensive study watersheds to detect changes in these concentrations and fluxes as a result of changes in atmospheric deposition. Atmospheric Hg not only enters watersheds through direct deposition, but also through uptake by vegetation. Mercury derived from atmospheric deposition that is directly deposited to surface waters or is transported along shallow flowpaths (see Figure 2.2) is expected to be more responsive to changes in atmospheric deposition than Hg that is taken up by vegetation and is incorporated into soil organic matter or that infiltrates into deep groundwater. Mercury entering a watershed by different mechanisms (i.e., direct deposition, vegetation uptake) or that is transported along different flowpaths is likely to have different potentials for either evasion back to the atmosphere or methylation. These factors must be considered in the design of a monitoring program to ascertain how watersheds respond to changes in atmospheric Hg deposition. We also recommend that the program of measurements at intensive study watersheds be integrated with a program or programs to assess the spatial extent of changes in atmospheric Hg deposition. This spatial assessment might be done with forest floor surveys and surface water surveys (see Chapter 3) that are population based and conducted at regular intervals (e.g., every 5 years).

New Hg from recent atmospheric deposition will combine in soil with both native mineral Hg and Hg associated with historical deposition. This pool of combined Hg may leach into groundwater and surface waters. Depending on the characteristics of the watershed, monitoring studies need to not only evaluate the potential for leaching of this soil Hg, but also be able to determine the incremental effects associated with changes in air deposition. Because soils may have large pools of historically deposited and mineral Hg (relative to the contribution from new deposition), monitoring studies alone may not be able to apportion the THg and MeHg loads, between historical or native Hg and new Hg deposition. Such studies could be facilitated by the use of stable isotope tracers (e.g., Hintelmann et al. 2002).

Mercury in soil is not only likely to have a different potential for evasion and methylation than Hg in runoff, but soil Hg may be perturbed by land disturbance. Land disturbances that are particularly relevant to Hg cycling include the formation of wetlands and flooding of reservoirs (Rudd 1995; see Chapter 3). Disturbances such as clear-cutting can also result in marked increases in the release of THg and MeHg from soils (Munthe and Hultberg 2003; Porvari et al. 2003). Fire can result in large Hg losses by volatilization (Grigal 2002).

The response of surface water Hg to changes in atmospheric Hg deposition will be influenced by the existing Hg pools in soil and by terrestrial processes that modify the transport of Hg deposition to surface waters (e.g., adsorption, vegetation uptake, mineralization, reduction, and evasion). To discern any change in loading to watersheds, multi-year studies will be necessary to detect real trends in the response of surface waters. Additionally, watersheds with low pools of native mineral Hg should be chosen for study. Because the potential for Hg to methylate varies considerably and can be influenced by a myriad of factors other than Hg loading, studies to ascertain changes in the formation and mass transfer of MeHg in response to changes in Hg deposition will be important.

A number of research projects on the dynamics of THg and MeHg in forested catchments and wetlands have been conducted in Sweden (Iverfeldt 1991; Hultberg et al. 1994; Munthe et al. 1995a, 1995b; Lee et al. 1994, 2000); North America (St. Louis et al. 1996; Driscoll et al. 1998); and Germany (Schwesig et al. 1999; Schwesig and Matzner 2000). Most of these studies have focused on input/output budgets and relationships between THg and MeHg behavior and hydrology, and interactions with other solutes (e.g., DOC). A summary of input and output fluxes of THg in 9 catchments in Europe and the United States and Canada is presented in Figure 2.8. Although the catchment characteristics are variable, some common features can be found. For example, THg and MeHg inputs (i.e., throughfall and litterfall) greatly exceed wet deposition. Also for all sites, inputs of THg greatly exceed drainage losses.

2.3.2 INTENSIVE WATERSHED MONITORING

We propose that a relatively small number of watershed sites be established for intensive monitoring. These sites should be selected based on the criteria for site selection discussed previously (see Section 2.1.3). In particular, these sites should be located in areas that:

1) Are sensitive to elevated atmospheric deposition of Hg
2) Are expected to experience marked changes in deposition in response to changes in Hg emissions
3) Represent background conditions (i.e., remote from local and regional sources of Hg and would represent changes in global emission of Hg(0))

Sensitive sites would include sites that are receiving high inputs of atmospheric Hg deposition and sites with aquatic ecosystems where top end predators have high levels of Hg. We also recommend that urban sites with elevated atmospheric Hg deposition and forest sites with shallow hydrologic flowpaths, wetlands, and unproductive aquatic ecosystems should strongly be considered as candidate sites.

These sites should be co-located with other Hg monitoring activities, including intensive air chemistry measurements (see Section 2.2.6), total ecosystem deposition (see Section 2.2.8), and comprehensive monitoring of adjacent aquatic ecosystems (e.g., lake, reservoir, estuary; water, and sediment chemistry (see Chapter 3), aquatic biota (see Chapter 4), and wildlife (see Chapter 5)). For this proposed program, we do not envision that new sites would be established. Rather, the intensive watershed Hg monitoring program would partner with existing intensive ecosystem study sites. Examples of existing intensive watershed networks include the National Science Foundation (NSF) Long-Term Ecological Research (LTER) program, the International Cooperative Programme on Integrated Monitoring of Air Pollution Effects on Ecosystems of the UNECE Convention on Long-Range Transboundary Air Pollution (http://www.environment.fi/default.asp?contentid=17110&lan=en), the U.S. Geological Survey Watershed Energy and Biogeochemical Budgets (WEBB program, http://water.usgs.gov/webb/), and the U.S. Park Service watershed program. Sites within these networks have an established infrastructure to conduct comprehensive ecosystem research. This infrastructure includes laboratory buildings, a permanent field staff, gauged watersheds, and long-term records of meteorological, hydrological, and ecological data. Relevant data sets that are routinely collected at these sites are summarized in Table 2.3. This infrastructure and the collection of supporting data would be invaluable to an intensive watershed Hg monitoring program. Such intensive study sites have been previously used to document long-term trends in air pollutants to forest ecosystems and to establish mechanisms responsible for these changes (e.g., Likens et al. 1996; Driscoll et al. 2001; Palmer et al. 2004) and test models which simulate the effects of air pollution on complex ecosystems (e.g., Aber et al. 2002; Gbondo-Tugbawa et al. 2001, 2002). These facilities would greatly decrease the cost of an intensive watershed Hg monitoring program, provide critical data that would help interpret trends, and support the testing of Hg cycling models.

It will be a challenge to select a small number of intensive sites that are "representative" of a region. Catchments situated in the same region may (and do) react differently to atmospheric changes (both climatological changes and atmospheric Hg deposition) and watershed disturbance. Thus, understanding the main causalities, response, and regional impact may be biased if the intensive study watersheds are not well represented.

At an intensive watershed Hg monitoring site, it is envisioned that THg and MeHg would be measured in ambient concentrations of atmospheric Hg species (i.e.,

TABLE 2.3
Summary of ecosystem measurements that are routinely made at intensive watershed study sites

Measurement	Sampling interval	Measurements made
Meteorology	Real-time, with hourly averaging	Wind direction and speed, air temperature, relative humidity, solar radiation, precipitation
Wet deposition	Weekly	Major solutes
Throughfall	Weekly	Major solutes
Litterfall	Seasonally	Mass, major nutrients
Soils	Once for characterization, forest floor at 5-year intervals	
Vegetation	5 years	Mass, major nutrients
Soil solutions	Quarterly	Major solutes
Groundwater	Quarterly	Major solutes
Hydrology	10 minutes	Temperature, discharge
Stream chemistry	Weekly	Suspended solids, major solutes

TABLE 2.4
Summary of collections and measurements of THg and MeHg that should be made at intensive watershed Hg monitoring sites

Measurement	Recommended interval of collection
Wet deposition	Weekly
Throughfall	Weekly
Litterfall	Monthly
Soils	Once to characterize pools
Soil solutions	Quarterly
Groundwater	Quarterly
Streamwater	Weekly

Hg(0), PHg, RGHg), wet deposition, throughfall, and litterfall, as discussed in the program to determine total ecosystem deposition (see Section 2.2.8). A summary of the measurements of Hg species that should be made in an intensive watershed Hg monitoring program is provided in Table 2.4. We envision that stream water measurements of total and dissolved THg and total and dissolved MeHg would also be made.

In addition to helping determine trends of Hg in ecosystems in response to changes in emissions of Hg, data from intensive watershed Hg monitoring would be critical to the interpretation of chemistry and biology data in downstream aquatic ecosystems (see Chapters 3, 4, and 5). Detailed Hg data from intensive watershed Hg monitoring sites would be available for the parameterization and testing of biogeochemical Hg cycling models (e.g., Hudson et al. 1994). These models will be important tools to help interpret the mechanisms responsible for changes in the transport, fate, and bioavailability of Hg in response to changes in Hg emissions and in making future projections on how complex ecosystems might respond to future changes in Hg emissions.

2.3.3 Soil Surveys

Mercury concentrations in forest organic soils may be good indicators of both current and long-term atmospheric inputs to catchments across large spatial scales. The forest floor is a strong sink for atmospheric deposition of trace metals to temperate and boreal forest ecosystems. A forest floor survey could be a good indicator of spatial patterns of total Hg deposition to forest ecosystems. Repeated surveys of forest floor samples could provide a quantitative understanding of changes in Hg deposition over a period of decades.

2.3.3.1 Forest Floor Surveys

The forest floor (i.e., the organic horizon overlying the mineral soil in forests) is an accumulator of trace metals. Researchers have conducted forest floor surveys to examine regional patterns in the deposition of trace metals and changes in the deposition of trace metals (Andresen et al. 1980; Herrick and Friedland 1990; Friedland et al. 1992). It is relatively easy to collect forest floor samples. Typically, 15×15-cm blocks are sampled, digested, and analyzed for trace metal content. Researchers have demonstrated regional and elevational patterns in trace metal content, which correspond to metal deposition patterns (Johnson et al. 1982). Moreover, in the case of lead, researchers have effectively documented decreases in deposition across a region (Friedland et al. 1992).

As inputs of THg appear to be strongly retained in the forest floor, we recommend a forest floor survey be conducted in areas that are receiving elevated Hg deposition and where it is expected that deposition will change markedly. In this site, permanent sampling sites would be established. We envision that forest floor samples would be resurveyed at appropriate time intervals (e.g., 10 years). A forest floor survey would help clarify current patterns of total Hg deposition and potentially quantify the response of forests to decreases in Hg emissions and deposition.

2.3.3.2 Surface Water Surveys

Cluster sites should be established for synoptic surface water surveys. This approach is discussed in detail in Chapter 3.

REFERENCES

Abbott ML, Susong DD, Krabbenhoft DP, Rood AS. 2002. Mercury deposition near an industrial emission source in the western U.S. and comparison to ISC3 model predictions. Water Air Soil Pollut 139:95–114.

Aber JD, Ollinger SV, Driscoll CT, Likens GE, Holmes RT, Freuder RJ, Goodale CL. 2002. Inorganic N losses from a forested ecosystem in response to physical, chemical, biotic and climatic perturbances. Ecosystems 5:648–658.

Andresen AM, Johnson AH, Siccama TG. 1980. Levels of lead, copper and zinc in the forest floor in the northeastern U.S. J Environ Qual 9:293–296.

Arriya PA, Khalizov A, Gidas A. 2002. Reactions of gaseous mercury with atomic and molecular halogens: kinetics, product studies, and atmospheric implications. J Phys Chem A(106):7310–7320.

Bloom NS, Watras CJ. 1989. Observations of methylmercury in precipitation. Sci Total Environ 87/88:191–207.

Brosset C, Lord E. 1991. Mercury in precipitation and ambient air: a new scenario. Water Air Soil Pollut 56:493–506.

Bullock Jr, OR, Brehme KA. 2002. Atmospheric mercury simulation using the CMAQ model: formulation description and analysis of wet deposition results. Atmos Environ 36:2135–2146.

Driscoll CT, Otton JK, Iverfeldt A. 1994. Trace metals speciation and cycling. In: Moldan B, Cerny J, editors, Biogeochemistry of small catchments: a tool for environmental research. Chichester, England: J Wiley & Sons, p. 299–322.

Driscoll CT, Holsapple J, Schofield CL, Munson R. 1998. The chemistry and transport of mercury in a small wetland in the Adirondack region of New York, USA. Biogeochemistry 40:137–146.

Driscoll CT, Lawrence GB, Bulger AJ, Butler TJ, Cronan CS, Eagar C, Lambert KF, Likens GE, Stoddard JL, Weathers KC. 2001. Acidic deposition in the northeastern U.S.: sources and inputs, ecosystems effects, and management strategies. BioScience 51:180–198.

Ebinghaus R, Jennings SG, Schroeder WH, Berg T, Donaghy T, Guentzel J, Kenny C, Kock HH, Kvietkus K, Landing W, Muhleck T, Munthe J, Prestbo EM, Schneeberger D, Slemr F, Sommar J, Urba A, Wallschlager D, Xiao Z. 1999. International field intercomparison measurements of atmospheric mercury species at Mace Head, Ireland. Atmos Environ 18:3063–3073.

Ebinghaus R, Kock HH, Temme C, Einax JW, Lowe AG, Richter A, Burrows JP, Schroeder WH. 2002. Antarctic springtime depletion of atmospheric mercury. Environ Sci Technol 36(6):1238–1244.

Edner H, Faris GW, Sunesson A, Svanberg S. 1989. Atmospheric atomic mercury monitoring using differential adsorption LIDAR technique. Appl Opt 28:921.

Engstrom DR, Swain EB. 1997. Recent declines in atmospheric mercury deposition in the Upper Midwest. Environ Sci Technol 31(4):960–967.

Ericksen J, Gustin MS, Schorran D, Johnson D, Lindberg S, Coleman J. 2003. Accumulation of atmospheric mercury in forest foliage. Atmos Environ 37:1613–1622.

European Commission. 2001. Ambient Air Pollution by Mercury (Hg) - Position Paper on Mercury. European Commission Publisher, Office for Official Publications of the European Communities, Brussels, ISBN 92-894-2053-7.

Feng X, Sommar J, Gardfeldt K, Lindqvist O. 2000. Improved determination of gaseous divalent mercury in ambient air using KCl coated denuders. Fresenius' J Anal Chem 366(5):423–428.

Ferm M. 1979. Method for determination of atmospheric ammonia. Atmos Environ 13:1385–1393.

Frescholtz T. 2002. Assessing the role of vegetation as sources and sinks of atmospheric Hg using quaking aspen. MS thesis, University of Nevada Reno, 67 p.

Friedland AJ, Craig BW, Miller EK, Herrick GT, Siccama TG, Johnson AH. 1992. Decreasing lead levels in the forest floor of the northeastern USA. Ambio 21:400–403.

Gårdfeldt K, Jonsson M. 2003. Is bimolecular reduction of Hg (II) complexes possible in aqueous systems of environmental importance? J Phys Chem A(107):4478–4482.

Gbondo-Tugbawa SS, Driscoll CT, Aber JD, Likens GE. 2001. Evaluation of an integrated biogeochemical model (PnET-BGC) at a northern hardwood forest ecosystem. Water Resour Res 37:1057–1070.

Gbondo-Tugbawa SS, Driscoll CT, Mitchell MJ, Aber JD, Likens GE. 2002. A model to simulate the response of a northern hardwood forest ecosystem to changes in S deposition. Ecol Appl 12:8–23.

Grigal DF. 2002. Inputs and outputs of mercury from terrestrial watersheds: a review. Environ Rev 10:1–39.

Hanson PJ, Lindberg SE, Tabberer TA, Owens JG, Kim KH. 1995. Foliar exchange of mercury vapor: evidence for a compensation point. Water Air Soil Pollut 80:373–382.

Hedgecock IM, Pirrone N. 2004. Chasing quicksilver: modeling the atmospheric lifetime of $Hg^0_{(g)}$ in the marine boundary layer at various latitudes. Environ Sci Technol 38:69–76.

Herrick GT, Friedland AJ. 1990. Patterns of trace metal concentrations and acidity in montane forest soils of the northeastern United States. Water Air Soil Pollut 53:151–157.

Hicks BB, Baldocchi DD, Meyers TP, Hosker Jr RP, Matt DR. 1987. A preliminary multiple resistance routine for deriving deposition velocities from measured quantities. Water Air Soil Pollut 36:311–330.

Hicks BB, Hosker Jr RP, Meyers TP, Womack JD. 1991. Dry deposition inferential measurement techniques. I. Design and tests of a prototype meteorological and chemical system for determining dry deposition. Atmos Environ 25A:2345–2359.

Hintelmann H, St. Louis V, Scott K, Rudd J, Lindberg SE, Krabbenhoft D, Kelly C, Heyes A, Harris R, Hurley J. 2002. Reactivity and mobility of newly deposited mercury in a Boreal catchment. Environ Sci Technol 36:5034–5040.

Hrabik TR, Watras CJ. 2002. Recent declines in mercury concentration in a freshwater fishery: isolating the effects of de-acidification and decreased atmospheric mercury deposition in Little Rock Lake. Sci Total Environ 297:229–237.

Hudson RJM, Gherini SA, Watras CJ, Porcella DB. 1994. Modeling the biogeochemical cycle of mercury in lakes: the Mercury Cycling Model (MCM) and its application to the MTL study lakes. In: Watras CJ, Huckabee JW, editors, Mercury pollution integration and synthesis. Boca Raton (FL): Lewis Publishers, CRC Press Inc., p. 473–523.

Hultberg H, Iverfeldt Å, Lee Y-H. 1994. Methylmercury input/output and accumulation in forested catchments and critical loads for lakes in Southwestern Sweden. In: Watras CJ, Huckabee. JW, editors, Mercury pollution, integration and synthesis. Boca Raton (FL): Lewis Publishers, CRC Press, Inc.

Iverfeldt A. 1991. Occurrence and turnover of atmospheric mercury over the Nordic countries. Water Air Soil Pollut 56:251–265.

Johnson AH, Siccama TG, Friedland AJ. 1982. Spatial and temporal patterns of lead accumulation in the forest floor in the northeastern United States. J Environ Qual 11:577–580.

Johnson DW, Lindberg SE. 1995. Sources, sinks, and cycling of mercury in forested ecosystems. Water Air Soil Pollut 80:1069–1077.

Kamman NC, Lorey PM, Driscoll CT, Estabrook R, Major A, Pientka B, Glassford E. 2003. Assessment of mercury in waters, sediments, and biota of New Hampshire and Vermont lakes, USA, sampled using a geographically randomized design. Environ Toxicol Chem 23:1172–1186.

Keeler GJ, Glinsorn G, Pirrone N. 1995. Particulate mercury in the atmosphere: its significance, transport, transformation and sources. Water Air Soil Pollut 80:159–168.

Lalonde JD, Poulain AJ, Amyot M. 2002. The role of mercury redox reactions in snow on snow-to-air mercury transfer. Environ Sci Technol 36:174–178.

Landers DH, Overton WS, Linthurst RA, Brakke DF. 1988. Eastern Lake Survey: regional estimates of lake chemistry. Environ Sci Technol 22:128–135.

Landis MS, Vette AF, Keeler GJ. 2002. Atmospheric mercury in the Lake Michigan Basin: influence of the Chicago/Gary urban area. Environ Sci Technol 36(13):3000–3009.

Larjava K, Laitinen T, Vahlman T, Artmann S, Siemens V, Broekaert JAC, Klockow D. 1992. Measurements and control of mercury species in flue gases from liquid waste incineration. Int J Anal Chem 149:73–85.

Lee Y-H, Borg GCh, Iverfeldt Å, Hultberg H. 1994. Fluxes and turnover of methylmercury: mercury pools in forest soils. In: Watras CJ, Huckabee JW, editors, Mercury pollution, integration and synthesis. Boca Raton (FL): Lewis Publishers, CRC Press, Inc.

Lee YH, Bishop K, Munthe J. 2000. Do concepts about catchment cycling of methylmercury and mercury in boreal catchments stand the test of time? Six years of atmospheric inputs and runoff export at Svartberget, northern Sweden. Sci Total Environ 260:11–20.

Lee YH, Wängberg I, Munthe J. 2003. Sampling and analysis of gas-phase methylmercury in ambient air. Sci Total Environ 304:107–113.

Likens GE, Driscoll CT, Buso DC. 1996. Long-term effects of acid rain: response and recovery of a forest ecosystem. Science 272:244–246.

Lindberg SE, Garten Jr CT. 1988. Sources of sulfur in forest canopy throughfall. Nature 336:148–151.

Lindberg SE, Meyers TP, Taylor GE, Turner RR, Schroeder WH. 1992. Atmosphere/surface exchange of mercury in a forest: results of modeling and gradient approaches. J Geophys Res 97:2519–2528.

Lindberg SE. 1996. Forests and the global biogeochemical cycle of mercury: the importance of understanding air/vegetation exchange processes. In: Baeyens W, Ebinghaus R, Vasiliev O, editors, Global and regional mercury cycles: sources, fluxes and mass balances. NATO ASI Series, Vol. 21. Dordrecht, the Netherlands: Kluwer Academic Publishers, p. 359–380.

Lindberg SE, Stratton WJ. 1998. Atmospheric mercury speciation: concentrations and behavior of reactive gaseous mercury in ambient air. Environ Sci Technol 32:49–57.

Lindberg SE, Brooks SB, Lin C-J, Scott KJ, Landis MS, Stevens RK, Goodsite M, Richter A. 2002. The dynamic oxidation of gaseous mercury in the Arctic atmosphere at polar sunrise. Environ Sci Technol 36:1245–1256.

Lu JY, Schroeder WH, Berg T, Munthe J, Schneeberger D, Schaedlich F. 1998. A device for sampling and determination of total particulate mercury in ambient air. Anal Chem 70:2403–2408.

Lu JY, Schroeder WH. 1999. Sampling and determination of particulate mercury in ambient air: a review. Water Air Soil Pollut 112:279–295.

Mason RP, Abbott ML, Bodaly RA, Bullock Jr OR, Driscoll CT, Evers D, Lindberg SE, Murray M, Swain EB. 2005. Monitoring the response to changing mercury deposition. Environ Sci Technol 39:15–22A.

Meyers TP, Huebert BJ, Hicks BB. 1989. HNO_3 deposition to a deciduous forest. Boundary-Layer Meteorol 49:395–410.

Meyers TP, Hall ME, Lindberg SE. 1996. Use of the modified Bowen ratio technique to measure fluxes of trace gases. Atmos Environ 30:3321–3329.

Meyers TP, Finkelstein P, Clarke J, Ellestad T, Sims PF. 1998. A multilayer model for inferring dry deposition using standard meteorological measurements. J Geophys Res 103:22645–22661.

Munthe J, Xiao Z, Schroeder WH, Lindqvist O. 1991. Removal of gaseous mercury from air using a gold coated denuder. Atmos Environ 24A:2271–2274.

Munthe J, Hultberg H, Lee Y-H, Parkman H, Iverfeldt Å, Renberg I. 1995a. Trends of mercury and methylmercury in deposition, run-off water and sediments in relation to experimental manipulations and acidification. Water Air Soil Pollut 85(2):743–748.

Munthe J, Hultberg H, Iverfeldt Å. 1995b. Mechanisms of deposition of mercury and methylmercury to coniferous forests. Water Air Soil Pollut 80:363–371.

Munthe J, Hultberg H. 2004. Mercury and methylmercury in run-off from a forested catchment — concentrations, fluxes and their response to manipulations. Water Air Soil Pollut Focus 4:607–618.

Pai P, Karamchandani P, Seigneur C. 1997. Simulation of the regional atmospheric transport and fate of mercury using a comprehensive Eulerian model. Atmos Environ 31:2717–2732.

Palmer SM, Driscoll CT, Johnson CE. 2004. Long-term trends in soil solution and stream water chemistry at the Hubbard Brook Experimental Forest: relationship with landscape position. Biogeochemistry 68(1):51–70.

Pirrone N, Hedgecock I, Forlano L. 2000. The role of the ambient aerosol in the atmospheric processing of semi-volatile contaminants: a parameterised numerical model (GASPAR). J Geophys Res 105(D8):9773–9790.

Pirrone N, Costa P, Pacyna JM, Ferrara R. 2001. Atmospheric mercury emissions from anthropogenic sources in the Mediterranean region. Atmos Environ 35:2997–3006.

Pirrone N, Pacyna JM, Munthe J, Barth H. 2003a. Dynamic processes of mercury and other trace contaminants in the marine boundary layer of European seas — ELOISE II. Atmos Environ 37(S1):1–177.

Pirrone N, Ferrara R, Hedgecock IM, Kallos G, Mamane Y, Munthe J, Pacyna JM, Pytharoulis I, Sprovieri F, Voudouri A, Wangberg I. 2003b. Dynamic processes of mercury over the Mediterranean region: results from the Mediterranean Atmospheric Mercury Cycle System (MAMCS) project. Atmos Environ 37(S1):21–39.

Porvari P, Verta M, Munthe J, Haapanen M. 2003. Forestry practices increase mercury and methylmercury output from boreal forest catchments. Environ Sci Technol 37:2389–2393.

Possanzini M, Febo A, Liberti A. 1983. New design of a high-performance denuder for the sampling of atmospheric pollutants. Atmos Environ 17:2605–2610.

Rea AW, Lindberg SE, Keeler GJ. 2001. Dry deposition and foliar leaching of mercury and selected trace elements in deciduous forest throughfall. Atmos Environ 35:1352–2310.

Rudd JWM. 1995. Sources of methyl mercury to freshwater ecosystems: a review. Water Air Soil Pollut 80:697–713.

St. Louis VL, Rudd JWM, Kelly CA, Beaty KG, Flett RJ, Roulet NT. 1996. Production and loss of methylmercury and loss of total mercury from boreal forest catchments containing different types of wetlands. Environ Sci Technol 30:2719–2729.

St. Louis VL, Rudd JW, Kelly CA, Hall BD, Rolfhus KR, Scott KJ, Lindberg SE, Dong W. 2001. The importance of the forest canopy to fluxes of methyl mercury and total mercury to boreal ecosystems. Environ Sci Technol 35:3089–3098.

Schroeder WH, Anlauf KG, Barrie LA, Lu JY, Steffen A, Schneeberger DR, Berg T. 1998. Arctic springtime depletion of mercury. Nature 394:331–332.

Schroeder WH, Munthe J. 1998. Atmospheric mercury — an overview. Atmos Environ 32:809–822.

Schwesig D, Ilgen G, Matzner E. 1999. Mercury and methylmercury in upland and wetland acid forest soils of a watershed in NE-Bavaria, Germany. Water Air Soil Pollut 113:141–154.

Schwesig D, Matzner E. 2000. Pools and fluxes of mercury and methylmercury in two forested catchments in Germany. Sci Total Environ 260:213–223.

Slemr F, Brunke EG, Ebinghaus R, Temme C, Munthe J, Wängberg I, Schroeder WH, Steffen A, Berg T. 2003. Worldwide trend of atmospheric mercury since 1977. Geophys Res Lett 30:1516.

Sprovieri F, Pirrone N. 2000. A preliminary assessment of mercury levels in the Antarctic and Arctic troposphere. J Aerosol Sci 31:757–758.

Sprovieri F, Pirrone N, Hedgecock IM, Landis M, Stevens BK. 2002. Intensive atmospheric mercury measurements at Terra Nova Bay in Antarctica during November and December 2000. J Geophys Res 107(D23):4722–4729.

Sprovieri F, Pirrone N, Gårdfeldt K, Sommar J. 2003. Atmospheric mercury speciation in the marine boundary layer along 6000 km cruise path over the Mediterranean Sea. Atmos Environ 37(S1):63–72.

Stevens RK, Dzubay TG, Russwurm GM, Rickel D. 1978. Sampling and analysis of atmospheric sulfates and related species. Atmos Environ 12:55–68.

Stratton WJ, Lindberg SE, Perry CJ. 2001. Atmospheric mercury speciation: critical evaluation of a mist chamber method for measuring reactive gaseous mercury, Environ Sci Technol 35:170–177.

Susong DD, Abbott ML, Krabbenhoft DP. 2003. Mercury accumulation in snow on the Idaho National Engineering and Environmental Laboratory and surrounding region, southeast Idaho, USA. Environ Geol 43:357–363.

[USEPA] US Environmental Protection Agency. 1997. Mercury Study Report to Congress. Fate and Transport of Mercury in the Environment, Vol. III. EPA-452/R-97-005, US Environmental Protection Agency, US Government Printing Office, Washington, D.C.

[USEPA] US Environmental Protection Agency. 2003. Western Airborne Contaminants Assessment Program Research Plan, EPA/600/R-03/035, May 2003.

Wängberg I, Munthe J, Pirrone N, Iverfeldt Å, Bahlman E, Costa P, Ebinghaus R, Feng X, Ferrara R, Gårdfeldt K, Kock H, Lanzillotta E, Mamane Y, Mas F, Melamed E, Osnat Y, Prestbo E, Sommar J, Schmolke S, Spain G, Sprovieri F, Tuncel G. 2001. Atmospheric mercury distributions in North Europe and in the Mediterranean Region. Atmos Environ 35:3019–3025.

Wängberg I, Munthe J, Ebinghaus R, Gårdfeldt K, and Sommar J. 2003. Distribution of TPM in Northern Europe. Sci Total Environ 304:53–59.

Wilkening KE, Barrie LA, Engle M. 2000. Trans-Pacific air pollution. Science 290:65–67.

3 Monitoring and Evaluating Trends in Sediment and Water Indicators

David Krabbenhoft, Daniel Engstrom,
Cynthia Gilmour, Reed Harris, James Hurley,
and Robert Mason

ABSTRACT

As recently as a decade ago, a paucity of geographically dispersed and reliable data on mercury (Hg) and methylmercury (MeHg) in water and sediments would have made discussions of large-scale monitoring programs difficult to conceive or implement. Methodological advancements made over this time period, as well as substantial improvements in our overall scientific understanding of mercury sources, cycling and fate in the environment, have enabled scientists, land managers, and regulators to consider how environmental responses to changing mercury emissions and deposition could be monitored. A program whose ultimate goal is to assess environmental responses to changes in atmospheric Hg deposition will undoubtedly rely on sediment and water indicators as critical program components. For both water and sediment, a well established set of sampling protocols and analytical procedures will enable reliable data collection across a diverse set of aquatic ecosystems. Water-based indicators of Hg and MeHg have already been useful for documenting decadal-scale changes in Hg and MeHg concentrations in the Everglades of Florida and a seepage lake in northern Wisconsin. At both sites, changes in Hg deposition were also measured and linked to the environmental response. Unfortunately, there are very few other long-term records of Hg and MeHg in water and/or sediment, thus establishing widespread baselines or current trends is presently difficult. With increasing numbers of studies and monitoring efforts that utilized the collection of water and sediment samples, however, a growing database on Hg and MeHg is evolving that would be useful for site selection and establishing general contamination levels for a more coherent monitoring effort.

Within an aquatic ecosystem, water-based indicators are expected to be the first environmental compartment to respond to altered mercury loading and where change can be detected. The response would likely first manifest itself as a change in aqueous total Hg (HgT) concentration, and then later as a change in MeHg concentration. The MeHg/Hg ratio (also expressed as percent MeHg) is a measure of the efficiency of ecosystems to convert the load of inorganic Hg(II) into MeHg. Shifts in the value of this ratio could reflect changes in ecosystem conditions affecting methylmercury production or elimination other than Hg loading, thus helping to distinguish the effects of Hg loading from other confounding factors that can affect MeHg concentrations. Temporary changes in MeHg/Hg ratios could also reflect the time required for MeHg concentrations in ecosystems to respond to changes in Hg concentrations and methylation rates. These types of insights make the MeHg/Hg ratio a very useful indicator. In addition, a significant advantage to this indicator is that it requires no additional funding support, assuming Hg and MeHg measurements on sediment and water will be part of a routine monitoring plan.

Sediment-based indicators are also critically important for monitoring changes in Hg inputs to aquatic ecosystems, and are often better indicators (compared to water-based indicators) of changes to Hg loading that occur over several years to decades. Mercury researchers commonly sample sediments because they are good indicators of overall contamination levels, but also because near-surface sediments (<10 cm) are generally the most important site of MeHg formation in most ecosystems. Surficial sediment Hg and MeHg concentrations also drive most of the exchange with the overlying water column. The greatest challenge for using sediment as an indicator of change is deciding what depth interval of sediment current deposition is accumulating, as opposed to large relic pools that are deeper within sediments and likely have little influence on current contamination of aquatic food webs.

Sediment coring efforts have been a key area of research that has led to an improved understanding of historical changes and spatial gradients in Hg accumulation among lakes, reservoirs and bogs. Lakes are especially valuable for monitoring programs because they commonly yield the desired sediment accumulating characteristics to record changes, and because of their widespread occurrence. In addition, sediment accumulation rates of Hg are complimentary to direct monitoring of contemporary Hg concentrations in sediment, water, and biota because they provide a longer-term examination of the loading trend history for the monitoring site. Mercury accumulation rate studies should be an effective indicator for comparing aquatic ecosystems from differing geographic regions across the US, and repeat measurements would only have to be conducted about every 10 years.

Although many aspects of a Hg monitoring program can be debated, one aspect that should not be compromised is that to be effective, such a program will need to include multi-media sampling (air, water, sediment and biota) to document the causal factors and possible beneficial changes resulting from future Hg emission reductions. Highly coordinated sampling for the atmosphere, watersheds, and biota will be requisite to yield the most interpretable results that can reliably attribute change to the appropriate driving factors, and quantify environmental improvement.

3.1 INTRODUCTION

It is not clear from existing data sets whether Hg concentrations in water, sediments, and ultimately fish will respond over months, years, or decades following changes

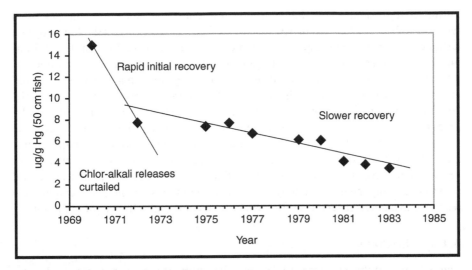

FIGURE 3.1 Observed mercury concentrations in standardized 50-cm walleye from Clay Lake, Ontario (1970–1983) following reductions in mercury releases from an upstream chlor-alkali facility. (Source: Data from Parks and Hamilton 1987.)

in atmospheric deposition. Based on our current understanding, response times over this entire range can be expected. Whether a system responds quickly or slowly to possible reductions in loading, however, will largely depend on its internal ability to remove Hg from actively cycling pools (i.e., Hg retirement). For example, in the 1960s and 1970s, abatement of point source Hg releases to many aquatic ecosystems led to rapid reductions in native fish Hg levels. However, as in the case of Clay Lake (Ontario, Canada), where Hg direct releases from a nearby chlor-alkali plant were eliminated, a rapid initial reduction in fish Hg levels can be followed by a prolonged slower recovery trend (Parks and Hamilton 1987, Figure 3.1). Whether the prolonged response is due to continued low-level releases from local point sources, recycling of relic contamination from sediments, or recent atmospheric deposition is not clear. However, lessons learned from some of these older studies will likely be useful for anticipating the timing and magnitude of responses from future Hg emission reductions. If the Hg reduction rate is large compared to the inventories of Hg in the ecosystem of interest, the response will likely be quicker and larger in magnitude. Thus, to anticipate environmental responses, we first need the ability to relate which Hg sources, inventories, and sinks are driving current conditions. To provide this understanding, researchers are presently working on identifying what forms of Hg are methylated and bioaccumulated, and whether newly deposited Hg behaves similarly to relic Hg in sediments (i.e., more or less reactive), and what depths of sediments (or soils) and water columns (epilimnetic vs. hypolimnetic) are involved in methylation and interact with other parts of the ecosystem. In the absence of knowing this information, monitoring efforts to reliably document responses to change could last years or decades, and the first few years of monitoring may not provide a good indication of the amount of time ultimately needed for water and sediment concentrations of Hg (regardless of the specific chemical form) to stabilize.

Furthermore, response dynamics may be different between water, sediments, and the food web (e.g., observed in new reservoirs by St. Louis et al. 2004). Any program designed to consider the response dynamics of total and methyl Hg in the environment should therefore consider the potential for different response dynamics in different components of the overall ecosystem. Lag times may also be observed among the various trophic levels of the food web. In new reservoirs, for example, Hg concentrations in top predatory fish can lag changes in water or sediments by several years, due to the time required for changing concentrations to cascade through the food web (Hydro Québec and Genivar 1997). Because compartments are linked with feedback mechanisms in real ecosystems, it will probably prove necessary to have information on the HgT and MeHg response trends in several ecosystem components, including water and sediments, to help understand the response dynamics observed in fish, which is the societally important endpoint.

3.1.1 OBJECTIVES

The objective of this chapter is to describe the utility of various sediment and water indicators that could be used for the purposes of quantifying the environmental benefit of possible future reductions in atmospheric Hg emissions, deposition, and bioaccumulation. This chapter focuses entirely on the collection, analysis, and interpretation of data derived from the analysis of water and sediment samples from aquatic ecosystems that are contaminated by atmospheric Hg deposition. Detecting and quantifying changes at sites previously contaminated by large, point-source loads will likely be much more challenging. Mercury cycling in the environment is notoriously complex, and as such it will be critically important to include coordinated sampling in time and space across all environmental media (air, water, sediments, biota). In addition, most successful Hg research programs rely heavily on the collection of related ancillary data (e.g., water chemistry, water levels, flow rates), which will also be critical to the overall success of any Hg monitoring program. Similar to most interdisciplinary data programs, the sum of the individual components are of far less value, and provide less insight, than when integrated multimedia data are presented in context together. In addition, although it is not discussed specifically, sediment and water Hg concentrations are commonly used as calibration targets for Hg cycling models, and as such, these data will also serve an important function should a large-scale modeling program come from this monitoring effort. For example, water- and sediment-based indicators were recently used to calibrate Hg cycling models for a pilot Hg total maximum daily load (TMDL) assessment (Atkeson et al. 2003).

3.2 SEDIMENT AND WATER INDICATORS

3.2.1 CRITERIA FOR SELECTING SEDIMENT AND WATER INDICATORS

Candidate sediment and water indicators were evaluated using criteria that assess whether the indicators are likely to demonstrate the environmental response to changes in external loading of Hg to aquatic ecosystems over anticipated time scales (decadal). The following 7 criteria were identified and used for evaluating the suitability of candidate indicators:

1) *Responsiveness.* One of the key considerations of any proposed indicator is whether it will demonstrate a detectable response to changes in Hg loading on relatively short time scales (decadal or less). In most atmospheric-Hg contaminated settings, annual atmospheric Hg mass loading is very small when compared to intact Hg pools in sediments, thus bringing into question whether changes in loading will be discernable above natural variability in the near to mid term (e.g., within a decade). In addition, Hg concentrations (any species) are generally low in water (less 10 ng/L; Weiner et al. 2003) and sediments (less than about 250 ng/g dry weight; Weiner et al. 2003). Thus, any indicator must be able to distinguish changes in external loading from recycling of existing pools, but at anticipated low concentrations.

2) *Comparability.* To assess trends, data for an indicator must allow for assessments in both time and space domains. Water and sediment samples can exhibit a high degree of natural variability in Hg species concentrations, which is due to natural heterogeneity and variations caused by differing sampling and analytical methodologies. To achieve maximum ability to detect trends, monitoring efforts must minimize variability caused by sampling methods. The ability to describe and implement strict sampling protocols will be a critical to the success of the monitoring program.

3) *Integration capacity.* Aquatic ecosystems will likely exhibit a significant degree of variability in response times to changing Hg loads. As such, an effective Hg monitoring program will need indicators that are responsive to ranges in time scales (months to decades).

4) *Understanding and knowledge of confounding factors.* Several factors not necessarily related to total mass loading of atmospheric Hg can affect the concentration and speciation of Hg in sediment and water. To correctly attribute trends in any indicator to actual changes in Hg deposition versus 1 of these confounding factors, it is essential to have a good understanding of what these factors are, and how they affect results from possible indicators. A more complete discussion of possible confounding factors of our chosen indicators is presented in Section 3.5.

5) *Ease of sampling and analytical reliability.* Twenty years ago, scientists could not reliably collect water samples for any Hg species without introducing substantial sampling contamination artifacts. Since then, reliable sampling and analytical protocols have been developed and widely accepted by the scientific community. These protocols allow for the collection of reproducible sample results with the ability to discern several Hg species and phase distributions. Although this area of research continues to pursue new sampling and analytical methods that expand our understanding, well-established and published methods that can be deployed on large geographic scales and under varying ecological conditions should be followed by this program.

6) *Availability of existing databases.* Many of the procedures for sampling and analyzing water and sediment samples for Hg and MeHg have been in place for 10 to 20 years, and as such the existence of databases that can be extended rather than initiated are now possible. At present, Hg deposition is variable spatially and temporally; thus, existing databases that can help describe ongoing trends for specific indicators at multiple monitoring locations would greatly benefit a Hg monitoring program.

7) *Cost concerns.* A large-scale, long-term, multifaceted Hg monitoring program will be expensive to initiate and sustain, but not out of proportion with the potential ecological and human health costs. The cost of implementation for each potential indicator should be carefully considered when making fiscally limited choices, especially in anticipation of cost limitations that will likely constrain the program and not allow for all the proposed indicators.

3.3 RECOMMENDED INDICATORS

The scientific understanding of Hg speciation in the environment, although far from complete, has increased considerably because of steadily improving analytical and field methods during the past 2 decades. Mercury exists in the environment in 3 oxidation states: Hg(0), Hg(I), and Hg(II). For each valence, many chemical forms (e.g., elemental Hg, inorganic Hg, monomethyl Hg, dimethyl Hg) and operationally defined fractions (e.g., reactive Hg, colloidal-bound Hg) can occur in the sediment and water phases. Operationally defined fractions are presently an active area of research that is leading to an increased understanding of what specific pools of Hg are participatory in important processes such as methylation. However, their applicability to a standardized monitoring effort is not clear, and as such they were not included in our consideration of candidate indicators. Also, some advanced processing methodologies (e.g., colloidal size separations) have greatly added to our overall understanding of the state of Hg in the environment; but due to significant post-sampling processing, they are not easily applicable to monitoring efforts. Dimethylmercury, although extremely toxic, has only been observed in the marine environment at very low concentrations (averaging 0.016 ng/L in the North Atlantic; Mason et al. 1998), but it has not been confirmed in fresh waters and thus was not considered as an indicator. Finally, although elemental Hg (Hg^0) is the dominant species in the atmosphere (>95%), in water it is almost always a relatively small fraction of total Hg in aqueous solution (<5%). In addition, Hg^0 can be a very unstable species in water, with rapid reoxidation potential, and shows strong diel (24-hour) concentration dependencies (Krabbenhoft et al. 1998b) that make it poor choice as an indicator.

Given the limitations associated with several of the possible Hg species in water and sediment, it was concluded that the most likely applicable Hg species were HgT and MeHg. However, 7 indicators were identified that are based on HgT and MeHg measurements on sediment and water samples (see Table 3.1) and discussed next in the context of the evaluation criteria.

TABLE 3.1
Recommended criteria for sediment and water indicators for monitoring responses to change in mercury loading

Criterion	Importance of criterion	Extent to which the indicator satisfies the criterion:						
		HgT in sediment (top 1–2 cm)	MeHg in sediment (top 1–2 cm)	Percent MeHg in sediment	Instantaneous methylation rate	Sedimentary accumulation rate of Hg	HgT in surface water	MeHg in surface water
Response to change on annual (top) and decadal (bottom) time scales	To quantify the environmental benefit to Hg load reductions	Low / High	Medium / High	Medium / High	High / Medium	Low / High	Medium / High	Medium / High
Comparability across sites and ecosystems	To document the utility of the indicator and its probable reliability for detecting change in mercury loading	Medium	Medium	High	Medium	High	Medium to High	Medium to High
Integrates variability in space and time	To facilitate the defensible interpretation of monitoring results on mercury	High	High	High	Low	Medium to High	High	Medium
Knowledge of confounding factors	To ensure knowledge of organismal attributes that can affect mercury concentration and complicate interpretation of results	High	Medium	Medium	Medium	High	Medium to High	Medium

TABLE 3.1 (continued)
Recommended criteria for sediment and water indicators for monitoring responses to change in mercury loading

Criterion	Importance of criterion	Extent to which the indicator satisfies the criterion:						
		HgT in sediment (top 1–2 cm)	MeHg in sediment (top 1–2 cm)	Percent MeHg in sediment	Instantaneous methylation rate	Sedimentary accumulation rate of Hg	HgT in surface water	MeHg in surface water
Ease of sample acquisition and processing (analysis)	To select biotic indicators that have broad spatial coverage in a regional, national, or multinational monitoring program	Medium to High	Medium to High	Medium to High	Low	Medium	Medium to High	Medium to High
Availability of existing databases	To select biotic indicators with a significant role in the trophic transfer of MeHg in aquatic food webs	Medium	Low	Low	Low	Medium to Low	Medium	Medium
Cost concerns (here, High implies there are substantial costs associated with the indicator)	Spatio-temporal variation in trophic position can confound and complicate interpretation of trends in mercury concentration in the indicator	Low	Medium	Medium	Medium to High	Low, if done every 5–10 years	Low	Medium

3.3.1 Sediment-Based Indicators

As opposed to surface water that can respond quickly to changes in loading, sediments generally serve as integrative measures (inputs over a few years to decades) of Hg loading and accumulation for a specific location. In addition, sediment is a common environmental matrix for assessments of contamination level and potential toxicity (Long et al. 1995). As such, sediment-based indicators are highly relevant for monitoring loading changes that occur and are sustained over several years. Net Hg accumulation in the sediments of water bodies is an integrative indicator of direct deposition to the water surface, plus Hg transported from the watershed from stream flow and groundwater discharge, and less what is lost to evasion, seepage to groundwater, and streamwater outflow (Krabbenhoft et al. 1995). Watershed retention of atmospherically deposited Hg commonly ranges from 50 to greater than 90%, with large forested watersheds generally retaining a higher fraction of deposited Hg (e.g., Krabbenhoft and Babiarz 1992; Krabbenhoft et al. 1995; Lee et al. 1995; St. Louis et al. 1996; Babiarz et al. 1998). Because Hg in sediments reflects watershed transport processes, it can be an indicator of land use patterns, as well as patterns of Hg deposition, through time and space. Because inorganic Hg in bulk sediments is the substrate for methylation (Benoit et al. 2003), the Hg concentration in these matrices is also a key parameter linking Hg deposition to MeHg production, and to bioaccumulation in food webs.

3.3.1.1 Total Hg Concentration in Sediment

In many settings that have sediment accumulating basins, total Hg concentration in sediment has been shown to change in response to changes to external Hg loading. Dated depth profiles of HgT in sediment cores clearly show changes in Hg accumulation rates over time that correlate well with documented Hg utilization and environmental releases (Wang and Driscoll 1995; Engstrom and Swain 1997). Thus, the top few centimeters of sediment in an aquatic ecosystem can be useful for monitoring recent Hg deposition conditions, or to show Hg deposition gradients among or within regions.

Total Hg is generally reported as nanograms (ng) of Hg per gram of sediment on a dry weight basis. A total Hg analysis on sediment includes all forms of Hg (both inorganic and organic species) that are present in the digestion solution after strong chemical oxidation and subsequent analyses by cold vapor purge and trap, and detection with atomic fluorescence (USEPA 1996; Olund et al. 2004). The inorganic Hg concentration can be calculated by difference if MeHg is measured on a sample split. However, because MeHg is generally a small fraction (<5%) of HgT in most aquatic sediments (often within the error of the measurement), HgT concentrations, rather than inorganic Hg, are generally reported.

Similar to most Hg sampling methods, sampling sediments and soils require care in avoiding contamination artifacts due to improper sample handling. However, because Hg concentrations are much higher in solid matrices than in water, if commonly accepted trace-metal protocols are used, substantial contamination artifacts should be exceedingly rare. Also, because sediment Hg concentration profiles show strong variability with depth, care must be taken to not mix the sample before the target sample sediment depth (top 1 to 2 cm) has been acquired. This often means careful hand sampling in shallow water (e.g., push cores or careful skimming of the surficial sediment) and deep-water coring procedures that minimize sample disturbance (e.g., push cores, freeze coring, gravity coring, box coring, piston coring) and sectioning the core when in a stable setting. Spatial heterogeneity is also a concern, and composites of multiple replicate samples are generally needed to account for natural sample heterogeneity. For HgT analysis in sediment, the analytical relative percent difference (RPD) can be as high as 10 to 20%. Nevertheless, spatial heterogeneity is generally larger than analytical variability.

One-time sampling of HgT concentration in bottom sediment is marginally useful as an indicator of Hg deposition to aquatic ecosystems, but can be a useful marker of changes to loading when sampled repeatedly using the same methodology. Absolute HgT concentrations in bottom sediment are, in part, a function of Hg loading, but are modified by other possible Hg sources to the water body (transport and retention processes within watersheds) and the sediment mass accumulation rate. For example, water bodies with substantial suspended particulate matter (e.g., eutrophic lakes, reservoirs with high sediment inputs) will often show dilution of Hg concentrations in bottom sediments relative to water bodies with relatively low sedimentation rates (e.g., oligotrophic lakes), although atmospheric deposition rates may be similar. Thus, care must be taken not to base inferences of Hg loading rates on concentration profiles alone, but rather sediment accumulation rates (see Section 3.3.6). For this indicator to be useful in the context of monitoring changes to loading, only the very top (1 to 2 cm) of sediment should be sampled with the least possible disturbance of the sediment water interface, and by using the same sampling depth throughout the monitoring program. In addition, considerations for confounding factors that could lead to changes in HgT concentration that are not necessarily related to atmospheric Hg deposition (changes to mass sedimentation rates and other Hg sources in the basin) are critically important to ensure the proper interpretation of the data.

The ability to detect differences in Hg concentration in sediment through space and time depends on the degree of natural heterogeneity, and on the number of samples that can reasonably be obtained. Unlike water, natural sample variability for sediments is generally much higher than analytical reproducibility. For most sediment, composites of multiple replicate samples are generally needed to reduce variability to acceptable levels, along with homogenization of samples prior to analysis. Analysis of Hg requires care and expertise. It is critical that laboratories providing analysis for Hg monitoring projects provide method validation prior to start-up, and participate in inter-laboratory calibrations of sampling, storage, and analysis techniques during the course of the project.

Although the primary intent of this monitoring program is to assess change at specific locations, comparisons of HgT concentrations in sediment are commonly made among sites to infer Hg loading differences. There are several factors to consider when making comparisons of HgT concentration in sediments across ecosystem types, including grain size and organic matter content. Differences in these factors among sites can lead to highly skewed HgT data sets, and make direct comparisons among varying sediment types problematic. Normalization to organic matter content, or an explicit measure of Hg accumulation rate (see discussion in Section 3.3.6 below), can aid with interpretations of differences in Hg concentration among sediment types, if needed. Given the above caveats, Hg concentrations in sediments of similar texture and chemical composition, and when sampled using the same technique and at the same interval, will be a useful component of a Hg monitoring program.

3.3.1.2 MeHg Concentration in Sediment

Although MeHg generally represents only a small fraction (usually less than 5%) of the HgT pool in sediments, a significant amount of current research focuses on its formation, cycling, bioaccumulation, and toxicity (Wiener et al. 2003). Increased attention on this 1 component of the HgT pool in the environment is due to its toxicity and the observation that greater than 95% of the Hg in edible fish tissues is MeHg (Bloom 1992), and thus is responsible for most of the exposure to wildlife and humans. Methylmercury concentration in sediment reflects the balance of MeHg inputs and outputs in sediments, including *de novo* methylation and demethylation. Despite the number of processes that can affect MeHg concentrations, MeHg concentration has been reasonably well correlated with measured isotopic tracer estimates of methylation potential in a number of systems, as demonstrated for several sites across the Florida Everglades (Figure 3.2). These strong correlations suggest that intact sedimentary MeHg concentrations primarily reflect the rate of recent MeHg production within sediment. This is an important observation, given the previously described link between HgT in surficial sediments and atmospheric deposition, which then may link sediment MeHg concentration to changes in Hg loading.

Methylmercury in sediment is a useful indicator to assess the net impact of all the factors that impact net methylation, including changing Hg load, changes to the net bioaccessibility of inorganic Hg, and changes in bacterial activity. Although there are many factors controlling net formation of MeHg in the environment, 2 important factors are the abundance and availability of inorganic Hg, which in turn is related to the atmospheric deposition rate. Thus, understanding the role of changing Hg loads to changes in MeHg concentration in sediment is critical for linking positive benefits of load reductions to reduced exposure. For example, in ecosystems with benthic-dominated food webs such as the Everglades, MeHg in surface sediments is a strong predictor of MeHg in biota (Figure 3.3).

Although MeHg concentration in sediment generally relates positively to HgT concentration, there is some question whether HgT in sediment is the primary controlling factor (Rudd et al. 1983; Henry et al. 1995; Hurley et al. 1998; Bloom et al. 1999), or whether a fraction of the HgT pool (e.g., recently deposited Hg, labile Hg, net zero charged Hg-ligand pairs) is the causal factor. To test these observations, some researchers have recently initiated in-field dosing experiments (Hintelmann et al. 2002; Krabbenhoft et al. 2004). These field experiments employ traceable stable Hg isotopes so that the possible confounding effects of relic Hg pools can be isolated from the experimentally applied Hg load. Results from experiments conducted at 4 different sites in the Florida Everglades clearly show a positive relationship between the amount of inorganic Hg added and the amount of MeHg

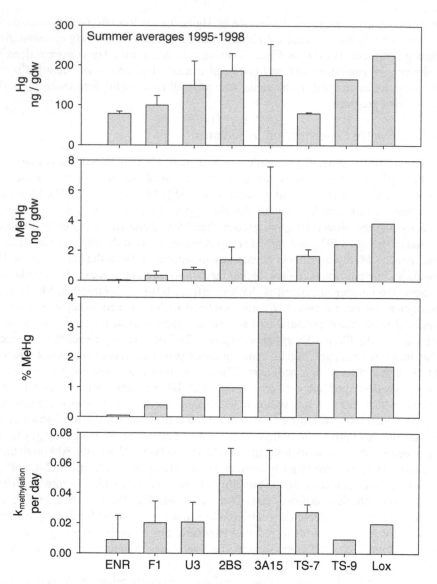

FIGURE 3.2 Comparison of HgT, MeHg, %MeHg, and estimated methylation rate for 8 sites across the Everglades (1995–1998). Each site was sampled 5 times over 4 years. At each time point, 5 separate cores were taken and analyzed, to assess variability and reduce standard error. The depth of soil sampling was 4 cm, assessed through prior analysis of depth profiles. In this wetland, a layer of flocculent material overlays the peat, and it is in this layer that methylation is strongest. Consideration of the methylation potential of detrital layers is often important in designing sampling programs for sediments and wetlands.

produced in sediments (Figure 3.4). Although the slope of the response varied by almost a factor of 100 among the test sites, which may seem surprising given that all the tests were conducted within the same ecosystem, all the sites showed a positive

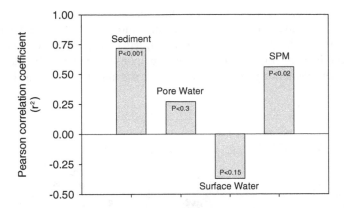

FIGURE 3.3 Pearson correlation coefficients between fish (Gambusia) Hg concentration and MeHg concentrations in various environmental media: sediment, porewater, surface water, and suspended particulate matter (SPM) from the Florida Everglades (1995–1998).

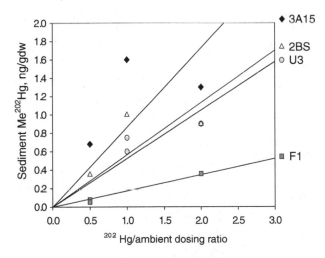

FIGURE 3.4 Results from the May 2000 dose-response experiment conducted *in situ* within mesocosms installed at 4 sites in the Florida Everglades and using isotopically labeled ^{202}Hg. Experimental conditions called for dosing at 0.5, 1.0, and 2.0 times the ambient loading rate of 22 ug/m^2/y.

relationship. Results from these mechanistic experiments and previous field research led us to conclude that we should expect to see positive correlations in sediment MeHg levels to changes in Hg loads.

Comparisons of MeHg and HgT sediment data from repeated sampling conducted at a specific location or within any single ecosystem appear to be relatively well-behaved and likely to be useful indicators. Comparisons of these sediment indicators among widely varying ecological settings, however, are less certain. Benoit et al. (2003) showed that HgT and MeHg data from a wide variety of aquatic ecosystems (rivers, estuaries, wetlands, and lakes), and that exhibit a larger range in HgT concentrations in sediment, result in a more complex (nonlinear) relation (Figure 3.5a). The large variation in the MeHg/HgT ratio observed from these data could reflect the real variability in ecological response represented by the far-ranging ecological settings among these study sites, or possibly the fact that these data are

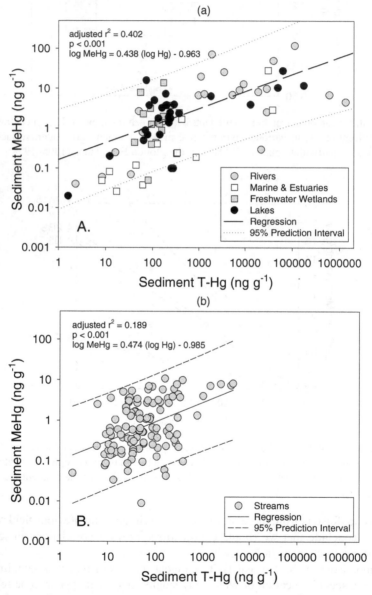

FIGURE 3.5 a) Relationship between HgT and MeHg in surface sediments across 49 ecosystems (from Benoit et al. 2003); and b) relationship between HgT and MeHg in surface sediments from 122 streams across the United States. (Source: From Krabbenhoft et al. 1999.)

derived from a variety of published sources that used variable sampling procedures and differing analytical laboratories. A similar relation is also derived for stream-bed sediment collected at 122 sites across the United States and using consistent sampling procedures and a single analytical lab (Figure 3.5b; Krabbenhoft et al. 1999). The striking similarity between these 2 data sets is somewhat surprising and supports the notion that MeHg will respond positively to changes in Hg loading and thus is a valuable indicator. It should be noted, however, that both data sets indicate that at any particular HgT concentration, almost a 2-order of magnitude range in MeHg concentration can be expected. So, although these data support the conclusion that reductions in HgT loading will lead to reductions in sediment MeHg concentrations, it will be difficult to *a priori* predict the absolute change in sediment MeHg concentration across a wide array of ecosystem types.

Heterogeneity of sediments is a major consideration when designing a monitoring program that includes sediment-based indicators. To illustrate the type of results that could be anticipated from a monitoring program, and to provide information on expected natural variability and ability to detect change, data on sediment HgT and MeHg for an extensively monitored ecosystem are shown in Figure 3.6. Lake 658 is the study lake for the METAALICUS project, a whole ecosystem Hg loading experiment (Hintelmann et al. 2002). Sediment texture and accumulation rates are relatively consistent throughout the basin. Repeated sampling of the top 2 cm of sediment at 0.5, 2, 4, and 6 m water depth throughout the ice-free season of 2001 showed obvious trends in measured concentrations of HgT and MeHg. The 0 to 2 cm sampling interval was chosen to represent the zone of maximum MeHg production, based on sediment depth profiles examined in 2000, the year before Hg loading was initiated. Multiple replicate sediment cores (>3) were taken at each time point, and care was taken to preserve depth gradients and to sample the top 2 cm accurately. For HgT, the calculated relative percent deviation (RPD) for within site variability is 16%, while site-to-site variability is about twice this amount. Spatial variability in MeHg is slightly higher, both within and among sites.

A second example from the Florida Everglades illustrates the importance of within-ecosystem variations in the natural sediment heterogeneity and the critical nature of this factor for using sediment indicators for detecting change. It should be noted that wetlands, with their heterogeneous root structures, probably offer a worst-case scenario of sediment MeHg variability. For this study, repeat sampling at 5 sites across the Everglades was conducted in which 5 replicate samples were collected on 5 separate occasions over the course of 4 years. The results show that sediment heterogeneity varies markedly among sites, and although it generally scales with increasing concentration, this is not necessarily a reliable predictor (Figure 3.7). It appears that patchiness of net MeHg production varies among these sampling sites, with the greatest variability observed where mean MeHg is the highest, and correspondingly less variability is associated with lower mean MeHg concentration. Overall, the mean RPD for sediment MeHg among all these sites is 53%.

As with HgT, concentration profiles for MeHg in sediment often show dramatic changes with depth and considerable spatial variability. Typically, maximum concentrations are observed at or near the oxic/anoxic interface, which is generally near

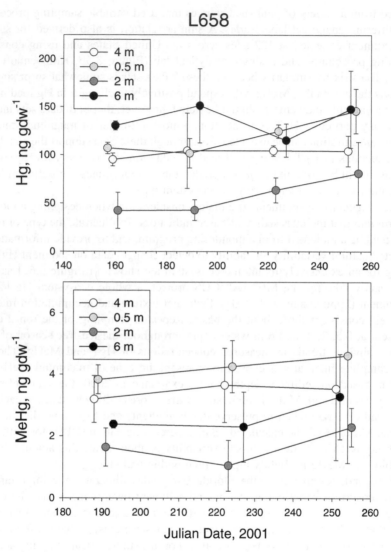

FIGURE 3.6 Measured concentrations of Hg and MeHg in the top 2 cm of sediments through time in 2001, at 4 discrete sediment sampling sites within Lake 658, the study lake of the METAALICUS project.

(within a few centimeters) the sediment/water interface. In some settings, the MeHg maxima can be much deeper in the profile (e.g., in emergent wetlands with fluctuating depth to water table and near root rhizomes). Selection of sampling depth is a critical part of MeHg sampling design. Prior to choosing a sampling depth, the zone of maximum MeHg production should be checked via depth profiles of MeHg concentration. Sampling depth should be selected based on the depth of the zone of high MeHg concentration.

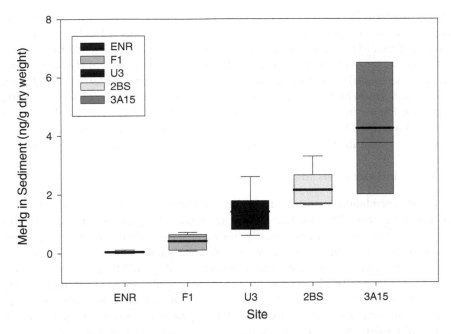

FIGURE 3.7 MeHg concentration (ng/g dry weight) in sediment from 5 sites in the Florida Everglades. Box plots represent 5 replicate samples taken at 4 different times over 4 years.

3.3.1.3 Percent MeHg in Sediment

Recent reviews on Hg methylation (e.g., Weiner et al. 2003; Benoit et al. 2003) suggest that MeHg abundance in the environment is enormously complex, and is affected by a number of factors, including many unrelated to Hg loading (see Section 3.6). In light of this, 1 simple, no added cost (assuming HgT and MeHg concentrations are measured as part of this program) method to help test whether possible trends in MeHg concentrations are related to changes in Hg loading is to normalize MeHg concentration to the HgT concentration of the same sample. This value is sometimes referred to as the percentage of HgT as MeHg (%MeHg), or the MeHg/HgT ratio.

Results from the Aquatic Cycling of Mercury in the Everglades (ACME) project provides an example of the use of this indicator, and highlights the importance of using sediment-based indicators as keys for monitoring net MeHg production over several years. Figure 3.2 shows data for HgT, MeHg, and %MeHg in shallow sediments across a north-to-south transect of the Florida Everglades. Total Hg (HgT) concentrations across these sites vary by only about a factor of 2, while MeHg varies by almost 2 orders of magnitude. The %MeHg shows an obvious maximum value in the central Everglades, which can be viewed as the location where the inorganic pool of Hg in sediment is the most bioavailable to the methylation process. The strong similarity between the %MeHg and the measured methylation rate constants supports this conclusion. This data set illustrates the utility of %MeHg in sediments as an indicator of MeHg production. It also shows how the %MeHg and methylation rate measurements can be used to factor out (normalize) the effect of HgT abundance in sediment on MeHg production. Because changes in sediment MeHg concentration are, in part, driven by HgT concentration, the %MeHg indicator will likely be a good indicator for linking possible changes in loading to possible changes observed in the food web. Finally, due to the low (or no) additional cost of utilizing this indicator, and the potential insights it offers, we recommend its use in monitoring efforts.

3.3.1.4   Instantaneous Methylation Rate

Although correlations between MeHg concentration and estimated methylation rate potential are generally good (suggesting that intact MeHg pools are generally produced *in situ*), this is not necessarily always the case (Benoit et al. 2003). Thus, instantaneous methylation rate assays on fresh sediments are useful to distinguish the influence of overall microbial activity (most notably, sulfate reduction) from the effects of sediment HgT concentration in governing the ambient sediment MeHg concentration. Direct measurement of methylation and demethylation rates also provides information about the location of these processes within ecosystems, which is useful for determining where the focus of monitoring efforts should be.

The measurement of potential Hg methylation and MeHg demethylation is significantly more complex than the measurement of HgT and MeHg concentrations in ambient samples. Methodological considerations include the maintenance of redox and temperature condition of samples during measurement, an understanding of the time course of both processes, and an understanding of the impact of spike level on the methylation and demethylation rate constants. Measurement of methylation and demethylation also requires the use of a tracer, and the ability to measure that trace analytically.

Estimates for the potential rates of methylation and demethylation are made by injecting dissolved Hg and MeHg spikes into intact sediment cores (Furutani and Rudd 1980; Gilmour and Riedel 1995; Hintelmann and Evans 1997), followed by short incubations (minutes to hours). Radiotracers (e.g., ^{203}Hg, ^{14}CH$_3$Hg) or stable Hg isotopes (e.g., ^{200}Hg, CH$_3$^{198}Hg) have been used by researchers in the past, and the spike concentrations can be close to tracer levels. Potential methylation rates are estimated by measuring the formation of the end-product (MeHg), while demethylation is measured through the loss of the methylated parent substrate. The rate estimates for these transformation processes can only be viewed as "potential rates" because the relative bioaccessibility of these introduced substrates compared to the natural inventories of inorganic Hg and MeHg is unknown, but both are probably more available to microbial communities. The demethylation product (inorganic Hg) is often difficult to measure against the large background of inorganic Hg in sediments and soils and, as a result, is often less precise.

Interpretation of methylation rate measurements can be complex because of the need to understand the time course of methylation/demethylation, and the dose-response to different levels of spiking, which can have a profound effect on the estimated rate constants. Many environmental factors can also influence estimates of these processes, and a more detailed discussion of these is provided in Section 3.5. In addition, although measurements of methylation and demethylation serve as indicators of microbial activity, they also reflect the kinetics of complexation of the Hg and MeHg spikes during the time of incubation. A number of studies have shown that the complexation of Hg spikes within hours of addition to sediments (even when pre-equilibrated with site water) is not the same as *in situ* Hg. Mercury spikes appear to be less strongly partitioned to sediments than are *in situ* Hg pools and thus unrealistically low rate estimates generally result from applying rate constants to porewater pools of Hg, and unrealistically high values come from assuming that the entire HgT pool in sediments is bioavailable (Krabbenhoft et al. 1998a). Nevertheless, MeHg concentrations in sediments and soils are often well correlated with instantaneous methylation rate estimates made from a relatively bioavailable Hg spike. This suggests that it is the most labile Hg that undergoes methylation *in situ*.

In the context of a monitoring program, methylation rate measurements would be part of a suite of process tools that would aid in the interpretation of whether changes in Hg loading, or possibly other confounding factors, are responsible for responses observed in other components of the monitoring program (e.g., aquatic biota). More specifically, rate measurements offer insights over and above MeHg concentration or the %MeHg by helping to assess changes in Hg bioaccessibility and microbial activity within and among aquatic ecosystems.

3.3.1.5 Sediment Hg Accumulation Rates in Dated Cores

Lake sediments, peat bogs, and ice cores have been used successfully for regional and global studies of modern and historical atmospheric Hg depositional patterns (e.g., Swain et al. 1992; Engstrom and Swain 1997; Benoit et al. 1998; Bindler et al. 2001; Lamborg et al. 2002; Schuster et al. 2002). Lake sediments are especially valuable because they occur over broad geographic regions. These natural archives are particularly well suited to examine the global/regional nature of atmospheric Hg dispersion and deposition, and are complimentary to direct monitoring of contemporary Hg concentrations in sediment, water, and biota. Lake-sediment records are particularly effective moderate-to-long (several years to centuries) trend indicators because 1) they smooth short-term variations in Hg deposition, 2) they integrate spatial variability in Hg flux to lakes and their catchments, 3) there is a large body of experimental and observational evidence for their reliability, and 4) there are well-established protocols for the collection, processing, and interpretation of sediment-core records.

Sediment Hg and MeHg accumulation rates in dated sediment cores are used to evaluate changes in the delivery of Hg to lakes through time, to compare the magnitude of change among lakes and regions, and to assess sediment burial rates for Hg in watershed mass-balance studies. Numerous studies have shown that sediment concentrations of HgT are relatively stable following burial and undergo little diagenetic remobilization (Fitzgerald et al. 1998), whereas more limited data on MeHg suggests substantial post-depositional losses through demethylation or diffusion to the overlying water. Additional work is needed to evaluate whether a modified signal for MeHg production can be retained in more deeply buried (older) sediments. A variety of studies employing dated sediment cores indicates that at a resolution of years to decades it is routinely achievable, thus making it possible to observe changes in sediment accumulation rates that are attributable to atmospheric loading changes at the same time scales.

The major difficulties in using lake sediments to track trends in Hg deposition involve the complexity of the sedimentary process. Well-behaved Hg records require conformable sediment burial that retains Hg in proportion to its load to the lake as well as the chronological markers used to date the core. Problems can arise when sediments are severely perturbed by slumping, mixing, or variability in sediment deposition across the lake bottom. These problems can often be avoided by the careful selection of study lakes and core sites, although natural variability in sediment deposition, which occurs in all lakes, can only be accommodated by collection of multiple cores. This is especially true when the signal strength for temporal change in Hg inputs is small — as might be the case for projected reductions in Hg deposition in the United States. For the purposes of a Hg monitoring program and documenting possible changes to recent Hg deposition, a minimum of 3 cores per lake is recommended. These cores should be collected in widely spaced locations across the profundal region of the basin and, as far as possible, from steep slopes or other lake-bottom irregularities. Similarity of timing, direction, and magnitude of change in Hg accumulation among cores is a robust indicator of temporal changes in Hg flux to the lake.

A secondary problem in interpreting Hg-sediment records is input of Hg from the catchment due to erosion (solid phase) and solubilization (aqueous phase) processes. Export of Hg from catchment soils to downstream aquatic systems can account for anywhere from <5% of a lake's Hg budget (seepage lakes) to >90% (drainage lakes with large catchments or high runoff yields). If catchment Hg inputs are substantial, the response of the sedimentary record to reduced direct (to the surface of the lake) atmospheric Hg deposition could be muted or significantly delayed by continued export from large inventories of Hg accumulated in soils (Kamman and Engstrom 2002). Moreover, catchment disturbances, both natural (fire, drought, beaver impoundment) and anthropogenic (logging, farming, urban development), can greatly alter the export of soil Hg to downstream lakes. For these reasons, it is essential that Hg-core records be obtained from multiple lakes (a minimum of 5) within a geographic cluster, both to reduce the likelihood of misinterpretation of trends not related to changes in Hg deposition and to explore the influence of catchment characteristics (size, land use) on response times to expected reductions in Hg deposition.

Detailed protocols for the collection and analysis of lake-sediment Hg records have already been published (EPRI 1996). The central elements include core collection and handling (sectioning), Hg analysis, and sediment dating. A large array of coring devices and approaches are documented in the paleo-limnological literature, and many (but not all) are suitable for recovering the undisturbed, high-resolution sediment profiles needed for this type of study. Piston coring, gravity coring, freeze coring, and diver-assisted (hand push) coring are all suitable under the correct circumstances. The main criteria include 1) the flocculent (high percentage of water) surficial sediments are recovered without disturbance; 2) a core of sufficient length (reaching back to pre-industrial times in most cases) is obtained; and 3) sediment displacement (compaction) is not appreciable. Core locations should be recorded precisely by GPS to allow re-coring for detection of subsequent trends in Hg deposition on a roughly 10-year re-occurring basis.

Cores are best extruded and sectioned on-site to avoid disturbing the flocculent sediments at the sediment/water interface, although careful transport (in upright, vertical position) to a laboratory for sectioning may be necessary under some circumstances. Accepted procedures to avoid sample contamination should be followed (acid-cleaned containers, gloves, etc.), although the typically high Hg content of lake sediments does not require the ultra-clean techniques necessary when sampling surface waters for Hg and MeHg. In all cases, the smeared edge of sediment on the surface of the core should be removed (trimmed away) during the extrusion/sampling process.

Sediment dating is one of the most critical and problematic aspects of obtaining reliable Hg-deposition records from lake sediments. Lead-210, the dating method of choice for obtaining a core chronology, has specialized analytical and interpretational procedures, and is briefly described here (Appleby 2001). The method relies on obtaining a detailed stratigraphic profile of ^{210}Pb activity (concentration) from the surface of the core to a depth at which a constant background (supported ^{210}Pb) is reached — typically 150 to 200 years. Lead-210 is measured by either alpha counting of ^{210}Po (a daughter isotope of ^{210}Pb) or direct gamma assay of ^{210}Pb. Alpha spectrometry methods have the advantage of higher precision and lower backgrounds, while gamma spectrometry provides a direct measure of supported ^{210}Pb (through ^{214}Pb) as well as ^{137}Cs, an ancillary dating tool. Neither method is inherently superior to the other.

Chronologies and sediment accumulation rates are derived from 1 of 2 simple models: 1) the constant rate of supply (c.r.s.) model, which assumes a constant flux of ^{210}Pb to the core-site but allows sediment input to vary; and 2) the cf:cs (constant flux:constant sedimentation) model that assumes both constant sedimentation and ^{210}Pb flux. For sites that have been substantially disturbed by human activity, sediment accumulation rates are almost never constant, and the c.r.s. model is required. Even remote lakes with little or no human disturbance often exhibit natural changes in sediment flux that require use of the c.r.s. model. Various modifications of these models to accommodate mixing (bioturbation) and other sedimentary processes can be applied, but almost always involve more complex assumptions, ancillary data, and independent dating markers. For a thorough treatment of dating models and their limitation, see Appleby (2001).

Finally, additional dating markers should be sought whenever possible to validate the ^{210}Pb chronology. This is especially critical for sites with disturbed watersheds and highly variable sedimentation rates, which are more prone to errors in ^{210}Pb dating. These ancillary dating tools include ^{137}Cs (for the 1964 peak in atmospheric nuclear testing), pollen indicators of historical land-use change (e.g., local European settlement), and profiles of other atmospheric contaminants with known input histories (e.g., pollution Pb and Pb isotopic ratios).

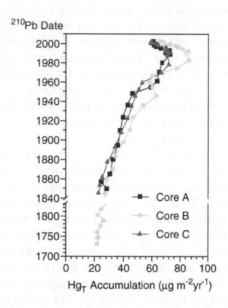

FIGURE 3.8 Historical trends in HgT accumulation in 3 sediment cores from Lake Annie, a seepage lake on the Archbold Biological Station, Highlands County, Florida. The cores were collected in 2003 and are from widely spaced locations within the profundal region of the basin. The similarity in trends among the cores reinforces the interpretation that the post-1850 increase and recent (post-1990) decline in Hg flux represent lake-wide changes in Hg loading. (*Source:* From Engstrom et al. 2003.)

The reconstruction of changes in atmospheric Hg deposition to lake sediments follows 3 general lines of interpretation: 1) temporal trends, 2) magnitude of change, and 3) whole-lake sedimentation rate. Temporal trends provide a qualitative assessment of the direction and timing of changes in Hg flux at a study site. Typically, Hg concentrations or accumulation rates are plotted against sediment age to give a visual sense of when Hg inputs began to rise (or fall), and the rate and magnitude of change. A typical example of this type of trend is shown in Figure 3.8, which shows 3 cores from Lake Annie, Florida (Engstrom et al. 2003). In this example, substantial increases in Hg accumulation over pre-development periods are observed, as well as more recent declines since the mid-1980s. Studies such as this are useful for supporting other observations, such as contemporaneous declines in fish and bird feather concentrations from the Everglades (Atkeson et al. 2003).

In situations where sediment accumulation rates have changed as the result of watershed disturbance, it is essential that Hg data be expressed as accumulation rates (rather than Hg concentrations). When multiple cores from multiple lakes all show the same trends in accumulation rates, it is likely driven by shifts in atmospheric deposition rather than changes in land use or lake processes, which are unlikely to be simultaneous among watersheds. The timing of the stratigraphic trends can be compared to estimates of historic changes in Hg emissions on local, regional, and global scales. Stratigraphic trends from some systems may not be in agreement with others in the region, which sometimes reveals within-watershed historical sources of Hg, such as past wastewater discharge or mining. However, within-watershed disturbances can often be corroborated from historical information (e.g., Balogh et al. 1999). Because of natural variability in sedimentation, limitations on stratigraphic (temporal) resolution, and mixing processes near the sediment/water interface, it is difficult to resolve from sediment records changes in Hg loading on time scales shorter than about a decade.

A comparison of Hg accumulation in a sediment core between reference (typically pre-industrial) and modern (or peak) conditions provides a more quantitative measure of the magnitude of change in atmospheric Hg deposition. This parameter is usually expressed as the ratio of recent (typically the past decade) to pre-industrial (pre-1850) Hg accumulation rates, with each time period represented by the average of several stratigraphic levels (Engstrom and Swain 1997). These flux ratios represent unit-less measures of changing Hg accumulation that are broadly comparable among sites and geographic regions. They are independent of individual site conditions that affect Hg sedimentation or atmospheric Hg loading (e.g., rainfall). These ratios provide robust measures of atmospheric impact if 1) local sources of Hg (e.g., Hg from direct discharge or local geologic sources) are minor, and 2) site conditions remain constant over the period of record.

Lake sediments are generally the primary sink for Hg inputs to lakes, and the determination of the whole-lake Hg sedimentation rate for mass balance calculations is a common objective of many lake studies. Because Hg concentrations and accumulation rates are spatially variable within a lake basin, Hg accumulation rates at a single core site cannot be easily extrapolated to the entire lake bottom. For reliable mass balance estimates, multiple dated cores (6 to 8 for small, bathymetrically simple basins) are required, with cores taken from representative areas of the lake bottom and spatially weighted to provide whole-lake burial rates over time. Core sites should be distributed in shallow (littoral) as well as profundal regions, and the portions of the lake bottom that do not accumulate fine-grained sediments (steep slopes, nearshore areas) also must be delineated. Dating precision and stratigraphic correlation limit the temporal resolution of whole-basin Hg accumulation to a decade or so for recent sediments (past half century) and 20 years or greater for older sediments.

3.3.2 WATER-BASED INDICATORS

When compared to sediments, the water column of a lake, reservoir, wetland, or stream generally represents a shorter-term indicator of Hg loading and processing relative to sediments or soils. Like sediments, however, the integration time of a water body can vary widely, depending on the hydrologic nature of the aquatic ecosystem and the species of Hg in water being considered. In some cases, and for some Hg species, surface waters may reflect changes in Hg loading and internal cycling over time periods of hours (e.g., reactions involving photochemical production of Hg(0) in shallow wetland systems; Krabbenhoft et al. 1998b) to several months to years (e.g., HgT in the water column, Great Lakes; Rolfhus et al. 2003). Although scientists have developed sampling and analytical strategies to quantify several forms of Hg in water (Gill and Bruland 1990; Krabbenhoft et al. 1998b), many forms (e.g., Hg(0) and reactive Hg) are short-lived or operationally defined and are not likely good indicators of changes in atmospheric Hg loading. As such, we limit our discussion here to HgT and MeHg in water, the forms that would likely have the most utility for reflecting changes in environmental conditions due to changes in atmospheric loading. Among surface waters, or within a given lake or stream, the abundances of MeHg and HgT can vary widely, and accurate measurement of their concentrations requires the steadfast application of trace-metal clean techniques to minimize sample contamination during collection, handling, and analysis, coupled with the application of highly sensitive analytical methods. When proper sample collection and preservation protocols are followed, inter-comparisons among laboratories that use accepted analytical methods for HgT and MeHg yield comparable results. When properly applied, water-based indicators are useful indicators of a robust monitoring program.

3.3.2.1   Total Hg in Water

Total Hg (HgT) in water is defined as the BrCl oxidized fraction of Hg (Bloom and Fitzgerald 1988; USEPA 1996). Over approximately the past 15 years, the research community has largely adopted this procedure for the analysis of HgT in water and, as a result, a wider geographic range of intercomparable data is available from the literature, and the expected range of concentrations in water is relatively well characterized. Unfortunately, long-term data sets of aqueous HgT concentrations from specific locations, upon which baselines and long-term variability can be ascertained, are rare. Total Hg in water can be further partitioned into dissolved (filter passing) and particulate phases (Gill and Bruland 1990), which is often useful for ascertaining sources within watersheds (Hurley et al. 1998). Even more sophisticated particle-separation techniques can be applied to surface water samples, such as those designed to assess the colloidal association of HgT in surface waters (Babiarz et al. 2001). These techniques can yield important information, such as the observation that a large portion of HgT draining forested and wetland watersheds is associated with colloids.

Concentrations of HgT in surface water represent a net integrative measure of the loading and removal rates for the water column of interest. Total Hg sources include direct atmospheric deposition, indirect deposition from watershed runoff, point sources, and internal recycling mechanisms such as sediment resuspension. Loss mechanisms for HgT in aquatic ecosystems include sedimentation, evasion, and riverine outflow. It is important to emphasize that the concentration of HgT for a given water body also depends strongly on other site or basin characteristics, such as water chemistry, land-use/land-cover characteristics, soil types, and hydrology. Because all these factors can have a controlling effect on aqueous Hg concentration, efforts aimed at monitoring and quantifying temporal changes must be attentive to the potential for the co-variation of these controlling factors. For example, a water body with rapidly increasing urbanization in its watershed would potentially not be useful for monitoring temporal changes due to presumed changes in atmospheric deposition. Carefully executed studies of spatial and temporal variations of HgT concentrations in surface water generally show well-behaved and predictable differences

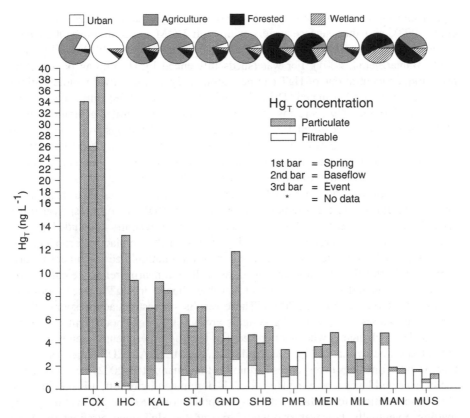

FIGURE 3.9 Mean concentration of HgT (filtered at 0.4 μM and particulate phases) in Lake Michigan tributaries ranked from highest to lowest, together with land use–land cover characteristics of watersheds. FOX = Fox River; IHC = Indiana Harbor Ship canal; KAL = Kalamazoo River; STJ = St Joseph River; GND = Grand River; SHB = Sheboygan River; PMR = Pere Marquette River; MEN = Menomonee River; MIL = Milwaukee River; MAN = Manistique River; MUS = Muskegon River.

among sites with varying settings. For example, HgT concentrations in tributaries to Lake Michigan vary both spatially and seasonally (Figure 3.9). Watersheds characterized by urban and agricultural land-use patterns generally have higher total Hg concentrations and greater portions on suspended particulates. In addition, the relative difference among these sites generally exceeds the seasonal differences in HgT concentration observed under high-flow versus low-flow conditions. Data such as these illustrate the improved likelihood of success for detecting temporal trends in aqueous HgT concentrations through repeat sampling at specific sites, as opposed to networks that use randomized site selection.

Because the principle source of Hg to most locations is atmospheric deposition from distant emission sources, concentrations of HgT in unfiltered water samples from lakes and streams lacking local anthropogenic or geologic sources are usually in the range of 0.3 to 8 ng/L (Hurley et al. 1995; Babiarz et al. 1998; Krabbenhoft et al. 1999). However, natural dissolved organic matter (DOM) readily complexes

Hg(II) and increases in stability and abundance in natural waters (Ravichandran et al. 1998), and thus surface waters with high DOM concentrations often have elevated (5 to 20 ng/L) HgT concentrations (Mierle and Ingram 1991; Hurley et al. 1995; Babiarz et al. 2001). For watersheds influenced by Hg mining or industrial pollution, concentrations of HgT can be significantly elevated over remote settings (about 10 to 40 ng/L; Wang and Driscoll 1995; Domagalski 1998; Hurley et al. 1998; Krabbenhoft et al. 1999). Streams draining areas with concentrated geologic sources of Hg or with large accumulations of Hg-contaminated tailings from Hg or gold mining can exceed 100 or even 1000 ng/L (Ganguli et al. 2000; Gray et al. 2000). Thus, mining and Hg-enriched areas would be poor choices for assessing changes in HgT concentration due to reductions in atmospheric loading.

It is not known how long it takes HgT concentrations in runoff to respond to changes in atmospheric Hg deposition, but the potential for long response times (decades or more) exists. In Sweden, researchers have recently concluded a long-term (exceeding 10 years) atmospheric Hg exclusion (by emplacing a roof) experiment for a 6300-m^2 catchment. By monitoring the manipulated catchment before, during, and after the roof was emplaced, as well as a nearby reference catchment, the authors concluded there was no observable change in HgT concentrations or fluxes (Munthe and Hultberg 2004). These results suggest decadal response times for similar watersheds, or flowing systems are possibly not good monitoring locations to detect change. A whole-watershed, Hg-loading experiment presently ongoing at the Experimental Lakes Area of Canada (i.e., the METAALICUS project) suggests that extended time lags to responses in runoff (Hintelmann et al. 2002) are likely. Thus, for streams and lakes where the Hg budget is dominated by runoff, it is possible that HgT concentrations for surface water may respond slowly to changes in atmospheric loading. Eventually, however, one would expect that HgT concentrations in runoff would dissipate once the leachable pool in forest soils is reduced. On the other hand, there are numerous other published reports showing that the HgT mass balance of some water bodies (e.g., seepage lakes) can be dominated by direct atmospheric inputs (Krabbenhoft et al. 1995; Watras et al. 1994; St. Louis et al. 1996). Under these conditions, it is reasonable to assume that HgT concentrations are likely to respond relatively rapidly to changes in loading, and show demonstrable changes within 1 to 5 years. With this in mind, monitoring programs should be designed considering the potential for different response dynamics in different aquatic ecosystem types (see Sections 3.4 and 3.6).

Total Hg concentration in water was the primary measurement in Hg cycling studies in the 1980s and early 1990s, and as such there is a relatively large body of data in the published literature that documents HgT concentrations in surface waters for the past 2 decades. That being said, however, long-term monitoring data at specific sites are few, and assessments of temporal trends (over many years) are rare. A water sample collected for a determination of HgT concentration represents a snapshot of conditions for a single receiving water at a particular time. In some rivers and reservoirs, HgT can vary annually by more than an order of magnitude due to flow conditions and variability in suspended particulates; thus, considerable caution should be exercised when choosing sampling sites and times. For lakes and wetlands that derive a greater proportion of their annual Hg mass budget from atmospheric deposition, variations will generally be less (Watras et al. 1994; Hurley et al. 1998). Thus, we can reasonably expect to see changes in HgT concentrations of seepage lakes due to changing loads over relatively short periods of time. However, non-loading related factors can also change the HgT concentration of surface waters, including pH, sediment loads, DOM, and eutrophication. An effective monitoring program needs to include these factors to effectively isolate changes in HgT that are due to loading changes (see Section 3.5 for a more complete discussion of these factors).

Examples of long-term (many years to decades) monitoring records of HgT concentration in water for specific locations are few. One such record exists for the central Everglades of Florida, where an 8-year record (1995 to 2003) HgT and MeHg in surface water has been assembled by the ACME project (Figure 3.10). The HgT concentration record shows considerable variation, but at the same time a declining overall trend is evident. When broken into thirds, the first third of this data string has a mean value of 1.9 ng/L, whereas the second and third segments have mean values of 1.3 and 1.2 ng/L, respectively; the 95% confidence intervals for these means (±0.34, 0.44, and 0.28 ng/L, respectively) support the assertion that there has been a decline in aqueous HgT concentration for this location. A regression analysis of the entire HgT data string indicates a negative slope (i.e., declining concentrations with time); however, the regression is not statistically significant (p = 0.16). This observation is intriguing because a recent analysis of long-term monitoring of piscivorous bird feathers and largemouth bass from the Everglades (Atkeson et al. 2003) concluded that Hg concentrations in biota have declined substantially in the past decade, which has been linked to declines in atmospheric deposition.

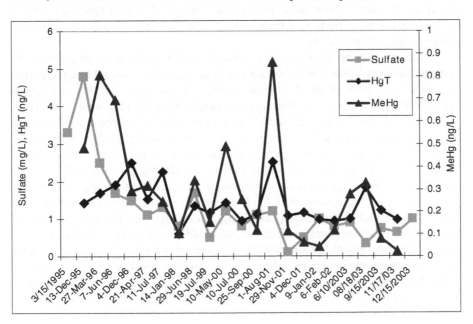

FIGURE 3.10 An 8-year record (1995–2003) of filtered HgT, MeHg, and sulfate from site 3A15 in the central Everglades.

Recently, Watras et al. (2000) concluded that a 0.04 ng/L/yr decline occurred in HgT concentration based on a 10-year monitoring period (1989–1999) for a single lake in northern Wisconsin, which resulted in an overall 39% decline in HgT. Although this average annual decline rate is modest and is at or near most reported analytical detection levels for HgT in water, when extended over several years, the decline in concentration is substantial. These authors conclude that the decline was driven by a 50% decline in Hg deposition rate from 1995 to 1999. This study represents a likely scenario of the magnitude of change in HgT concentration due to reduced Hg loading that may be realized by further US emission reductions, and reinforces the idea that many years of careful monitoring will be needed to definitely detect and quantify change.

Direct indications of how we might expect HgT concentrations in surface water to respond to changes in loading come from the METAALICUS project. During the first year of Hg additions during METAALICUS (2001), concentrations of inorganic Hg in the surface waters of Lake 658 nearly doubled as a result of nearly doubling the total load of Hg to the lake surface (atmospheric deposition plus runoff; H. Hintelmann, unpublished data). This trend was repeated in 2002 and 2003, suggesting that inorganic Hg concentrations in surface waters in small remote lakes respond quickly (within a year) to changes in the overall Hg loading to the lake. Exceptions would occur, however, in situations where a large component of the Hg load is on particles that settle quickly and do not release Hg to surface waters. What is not known, however, is whether the response to reductions will also be proportional or more complex in nature. Results from the METAALICUS project also illustrate that HgT concentrations in surface water (and presumably atmospheric deposition rates) are not static. Therefore, to establish current trends and/or baselines, considerable monitoring periods will be needed to establish true responses to possible future changes in emissions.

While monitoring HgT may not provide a clear and direct measure of the potential environmental benefit from reduced loading, it likely would be an important parameter of a multi-pronged monitoring strategy, the goal of which is to assess what factors drive changes to more critical monitoring endpoints, such as MeHg in water and fish tissue concentrations. Taken alone, the HgT concentration in water has been demonstrated to be an unreliable predictor of MeHg levels in water (Kelly et al. 1995), an outcome that is probably related to the many factors (beyond HgT concentration and loading) that control the bioaccessibility of inorganic Hg to food web uptake. While HgT alone does not give an indication of the bioaccessibility of Hg in an aquatic ecosystem, it is likely that HgT will be a critical measure that, when coupled with other environmental data, yields significant insights. For example, detailed HgT concentration data (including dissolved, particulate, and colloidal partitioning information), coupled with other related ancillary data and application of geochemical speciation modeling, could yield information on Hg speciation controls of methylation, and the role of aqueous HgT concentrations in regulating Hg levels in food webs. Thus, a successful monitoring program will likely need to include HgT concentrations in water from select receiving waters.

3.3.2.2 MeHg in Water

Methylmercury (MeHg) is the most toxic and bioaccumulative form of Hg in food webs (Wiener and Spry 1996; Wiener et al. 2003), and has been shown to be a good predictor of game fish Hg concentrations (Brumbaugh et al. 2001) but only comprises a small fraction of the total Hg pool in most surface waters (about 0.1 to 5%; Krabbenhoft et al. 1999; Wiener et al. 2003). So, although MeHg represents greater than 95% of the Hg in consumable game fish tissues (Bloom 1992) and commonly reaches levels of potential toxicological concern in remote locations (Brumbaugh et al. 2001; Wiener et al. 2003), most surface waters have MeHg concentrations ranging between about 0.04 and 0.8 ng/L (St. Louis et al. 1994; Hurley et al. 1995; Babiarz et al. 1998; Bodaly et al. 1998; Gilmour et al. 1998; Krabbenhoft et al. 1999; Waldron et al. 2000). Because of these extremely low aqueous concentrations, surface water sampling for MeHg must be executed with great care to avoid sampling artifacts (e.g., contamination from sample containers or improper handling of the sample, or accidental disturbance of the sediment/water interface causing sediment resuspension) and to ensure reliable analytical results (Olson and DeWild 1999; DeWild et al. 2002). Over the past 10 to 15 years, widespread acceptance of the need to practice trace-metal-clean protocols when collecting aqueous samples for HgT and MeHg has led to a much-improved ability to compare data generated by various research groups from widespread geographic areas, including the ability to examine for temporal trends. Continued strict adherence to these sampling and analytical protocols will be essential to any successful monitoring program.

Because MeHg is disproportionately important in driving the bioaccumulation of Hg, focused attention has been placed on understanding its occurrence, distribution, and mechanisms of formation. As a result, a much-improved understanding of MeHg formation in the environment has been realized over approximately the past 15 years (Benoit et al. 2003), and the complex array of factors that control its occurrence and distribution (e.g., availability of inorganic Hg, water and sediment biogeochemistry, activity of sulfate-reducing bacteria, microbial MeHg degradation rates, sulfur cycling, other metals, labile carbon substrates, anaerobic sediments, sediment-water exchange rates, photodegradation of MeHg, hydrology, and seasonal and inter-annual differences in all of the above). Thus, although MeHg concentration in water can be a good indicator of expected contamination levels of food webs, and is partly controlled by external Hg loading, all the factors influencing MeHg production need to be considered when the goal is to detect and quantify changes in environmental conditions directly attributable to reduced deposition rates. Monitoring efforts that include MeHg as an indicator, therefore, will necessarily need to include a more broad base of ancillary data (e.g., general water chemistry, hydrology, climate), as well as the indicators described in the accompanying chapters, in order to reliably ascribe the environmental benefits that are actually related to reductions in Hg deposition.

Like HgT, MeHg in water has several advantages over sediment-based indicators. First, MeHg in surface water is generally very dynamic, reflecting conditions over recent weeks to months, including recent changes to Hg loading. Second, compared

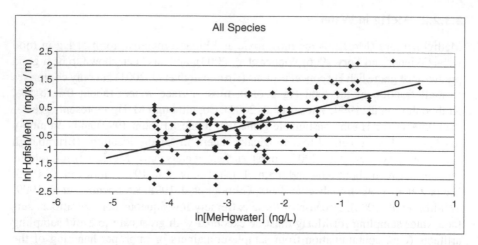

FIGURE 3.11 Relationship between aqueous MeHg on unfiltered water samples and fish (largemouth bass) Hg concentration normalized to length from streams across the United States. (*Source:* From Brumbaugh et al. 2001.)

to sediments, horizontal and vertical sample heterogeneity is relatively small in water columns, particularly filtered water samples; thus, field replication is generally better achieved with water samples. The dynamic character of MeHg in surface water does have drawbacks, however, and necessitates detailed attention be placed on recognizing the seasonal and spatial patterns that naturally arise. For example, maximal MeHg concentrations are commonly observed in hypolimnetic waters at the end of the summer and winter stratification periods for stratified lakes (Watras et al. 1994). It is critical for monitoring programs to conduct sampling efforts with these seasonal and spatial differences in mind. Finally, there is some evidence that aqueous MeHg concentration is a reliable indicator of fish Hg concentration (Figure 3.11). As such, state and federal monitoring programs may have interest in using this indicator as a substitute for much more expensive fish surveys.

Similar to HgT, continuous, long-term records of MeHg in surface water are rare. The 10-year record of HgT from a single northern Wisconsin lake discussed above (Watras et al. 2000) also had an accompanying MeHg data series. These authors suggest that the observed decline in HgT concentration (39%) from 1988 to 1999 was paired with a 53% decline in MeHg. Although the average annual rate of decline (0.004 ng MeHg/L/year) for this trend was beyond present analytical method detection limits, when monitored over a 10-year period, the trend was detectable and significant. Interestingly, the overall decline in MeHg concentration is more comparable to estimated change in atmospheric loading rate (50%) than the decline in HgT in surface water. Given the variability in HgT and MeHg concentrations in this data set, it is probably not possible to conclude that MeHg concentration is a more sensitive measure of change in atmospheric loading.

Methylmercury concentrations at site 3A-15 in the Florida Everglades have shown distinct declining trends over the 8-year period from 1995 to 2003. A regression analysis of this data string shows a statistically significant ($p = 0.048$) decline rate of about 0.043 ng/L/year, and an overall decline of about 0.35 ng/L over the 8-year period (70% decrease). These declines are similar in magnitude to the observed declines in fish and bird feather Hg concentrations reported by Atkeson et al. (2003) from the same region of the Everglades. It is interesting to note that in both Wisconsin and Florida, the observed declines in aqueous MeHg concentration are greater than the coincident declines in HgT. Whether or not the declines in environmental MeHg concentrations from the Everglades and Wisconsin can be directly attributable to commensurate changes in atmospheric deposition rates is not certain due to changes in other known contributing factors at both locations. Watras et al. (2000) concluded that surface water MeHg declines were at least partly driven by coincident changes in other known controlling factors (sulfate and DOM declines, and pH increase). In the case of the Everglades, atmospheric deposition rates of Hg in South Florida have been inferred to decline based on dated sediment cores (Engstrom et al. 2003). However, similar to the Wisconsin study, recent data from the monitoring site in the Everglades have revealed that substantial declines in sulfate concentration (greater than 90%) were coincident with MeHg declines (Figure 3.10). Thus, although 20 to 30% declines in Hg deposition appear to have occurred in South Florida from 1990 to 2000, large declines in sulfate to concentrations that are probably limiting sulfate reduction (less than 1 mg SO_4/L) likely played a key synergistic role in MeHg declines. These 2 cases illustrate the importance of including MeHg as an indicator of environmental response to changing Hg loads, but also the need to collect ancillary data to aid interpretations.

As with sediments, the %MeHg in water samples is sometimes used by researchers as an integrative measure of "methylation efficiency" or Hg sensitivity of an ecosystem (Kelly et al. 1997; Krabbenhoft et al. 1999; Benoit et al. 2003). The added value of using %MeHg over MeHg or HgT concentrations lies in attempts to normalize the variability in MeHg levels that are attributable to differences in HgT availability levels across spatial and temporal gradients. For example, in some aquatic ecosystems, the concentrations of HgT in surface water can show regular seasonal cycles, whether from changes to external loads (e.g., rainy vs. dry periods) or internal processes (e.g., resuspension of bed sediments during spring runoff). As such, the %MeHg tends to dampen the spatial and temporal variability exhibited in MeHg and HgT data alone, and is potentially a better indicator of changes to bioaccessibility of inorganic Hg for methylation. Recent research that employs the use of stable isotope tracers to document the effects of changing Hg loads (e.g., the METAALICUS project) has indicated that recent Hg additions are more available to methylating microbes than relic contamination (David Krabbenhoft, USGS, unpublished data). Therefore, the %MeHg may be an effective indicator of changes in the net bioaccessibility of the HgT pool that may not be discernable through HgT measurements alone.

Percent MeHg (%MeHg) is a relatively untested indicator of change in ecosystems. For the proposes of detecting and quantifying environmental change due to changes in external Hg loading, however, it would appear to be a useful tool. Given that MeHg and HgT are likely indicators for any Hg monitoring program, and thus there is little (or no) added cost associated, we recommend including it. For the previously discussed case of the Florida Everglades (Figure 3.10), a regression analysis of the %MeHg from 1995 to 2003 shows a decline of approximately 66%, and the regression is more significant (p = 0.012) than the observed declines in MeHg or HgT individually. Total Hg and MeHg are positively and significantly correlated (r^2 = 0.42; p = 0.003) at this site, which is not surprising given that inorganic Hg is 1 of the essential elements for the formation of MeHg. The %MeHg in surface water of the Everglades does show a generally declining trend. However, there are obvious spikes that are always timed with the onset of summer, which is also the high rainfall period and when the majority of "new" Hg deposition occurs. Based on this analysis, the %MeHg indicator not only appears to be a good indicator of the efficiency (or ability of an ecosystem to methylate) over both short and long time intervals.

3.4 MONITORING STRATEGY

The published literature indicates that Hg contamination of the environment occurs to some degree around the globe. Anticipating the current level of contamination, especially of indigenous food webs, remains difficult to predict due to the complex relationship that exists between Hg loading rates and top predator accumulation levels. Correspondingly, it will be equally difficult to anticipate the environmental response to possible reductions in Hg loading. Therefore, to quantify the environmental response to a change in Hg deposition at any location, monitoring data to set baselines and current trends is necessary before the change occurs. In addition, because sources of Hg in the United States originate both nationally and internationally, the network should be established to provide an estimate of whether observed changes are due to domestic or foreign actions. Thus, the network should be at least national (continental) in scale. Priority should be given to sites with rich existing databases and would seek to develop ongoing collaborations with other efforts that meet multiple needs (e.g., global change, urban sprawl, and changing land-use issues).

An Hg monitoring network that would have the greatest ability to detect change and quantify improvement in environmental Hg contamination levels would necessarily have good geographic coverage of a broad range of ecosystems (or ecosystem types). One possible network design employs the use of a series of grouped (geographically close) monitoring sites (or clusters) that are distributed across a range of ecoregions, and at which multimedia (water, sediment, and biota) monitoring would occur on a regular basis. We anticipate that clustered sites would be established within at least 3 and not more than 10 ecoregions. A limited number of sites within each cluster (probably 1 or 2) would be chosen as intensive study sites, where more detailed sampling would be conducted to provide an assessment of the driving factors behind any observed environmental responses.

Clustered sites would be located within a single ecoregion or at least have similar ecological characteristics (e.g., southeastern coastal plain streams), and would consist of a series of geographically proximal locations (approximately 10 to 20 water bodies) where similar atmospheric loads (or load reductions) are anticipated. Cluster sites would be sampled on a routine basis (e.g., quarterly) for a prolonged period of time. Individual sites within a cluster would be specifically chosen to represent a range of site characteristics that are known to affect Hg cycling (e.g., pH, DOC,

ANC, watershed/lake ratio). Site selection criteria for clusters would be based on multiple factors, including: geographic region, Hg deposition condition, watershed type, and water body type (e.g., lake, reservoir, river, and estuary). Presently, there are a number of candidate locations where clustered sites can be found that have the listed characteristics, including northern New England and the Adirondacks; the upper Midwest; the southeastern coastal plain; lakes in Ontario and Quebec (Canada); western mountain lakes, streams, and reservoirs, the arctic regions of Alaska; and several major coastal regions (San Francisco Bay/Delta, Chesapeake Bay, the Gulf Coast, and the Great Lakes).

The intensive monitoring site network would consist of approximately 20 sites with representation of each cluster (1 or 2 in each cluster). The intensive monitoring effort would include multimedia sampling (water, sediment, biota, and ancillary data such as flows, temperature, and general water chemistry) conducted about annually with the specific objective of assessing the driving causes for the environmental response to the change in Hg load. Priority would be placed on sites where maximum response is expected. These monitoring efforts should be highly coordinated with the watershed, atmospheric, and aquatic biota monitoring programs to achieve maximum interpretation of the data.

For the purposes of prescribing ecosystem-type considerations for a monitoring network, the goal(s) of the monitoring effort must first be defined. Certainly one goal of an Hg monitoring network could be to maximize the likelihood of detecting change. However, optimizing a network design for detecting change can be achieved by choosing aquatic ecosystem types that yield the least amount of natural variability in water and sediment characteristics, or by choosing sites where the greatest reductions in deposition can be expected to occur, or both. One example of this environmental setting would be an atmospherically dominated (mercury budget) seepage lake in a high-Hg deposition setting. On the other hand, another goal of a monitoring network could be to detect and quantify general benefits to environmental conditions that might result across a range of aquatic ecosystem types, or the United States more generally. In this case, sampling sites would include a variety of aquatic ecosystem types (e.g., lakes, wetlands, streams), and those where maximum range in changes might be expected or in areas that represent a maximum range in anticipated changes to atmospheric deposition. Site selection for a cluster in a particular ecoregion should involve careful consideration of ecosystem types that are representative or common to that area and potentially sensitive to Hg inputs (e.g., lowland streams in the coastal southeastern United States, soft-water lakes in the glaciated northern United States).

Although a preponderance of the founding research on Hg cycling in freshwater aquatic ecosystems has been on lakes and wetlands, it is not clear from the literature whether they are more or less "mercury sensitive" (i.e., prone to produce MeHg) and yield higher levels of MeHg in indigenous food webs than riverine ecosystems. Recent reviews do suggest that the density or abundance of wetlands in a basin is positively related to MeHg abundance (Wiener et al. 2003). It has not been clearly shown, however, whether this is a result of direct MeHg contributions from wetlands up-gradient from receiving waters, or secondary effects such greater DOC contributions from wetlands to lakes and streams, which may enhance bioaccessibility of inorganic Hg for methylation. Streams that have wide ranges in seasonal flow will likely exhibit wider ranges in unfiltered and filtered HgT and MeHg concentrations on an annual basis, and thus responses to changing Hg deposition at stream monitoring sites may not be as clear. Because no long-term Hg and MeHg records from streams are known to exist, it is uncertain whether they will respond more slowly or more quickly than still water bodies to changes in Hg atmospheric deposition; although it would seem reasonable that stream flow would be likely slower to respond, given that ties to terrestrial soil pools of Hg are closer compared to seepage lakes. The experimental watershed results from Sweden would indicate a long delayed response (Munthe and Hultberg 2004). However, because a substantial number of streams across the United States have posted fish consumption advisories for Hg, inclusion of stream sites would seem logical. In addition, streams and rivers transport MeHg to down-gradient ecosystems, such as coastal margins and estuaries, where presently very little is known about Hg and MeHg budgets. Obviously, a network designed to measure the general environmental benefit will be more costly (more sites with greater geographic distribution and probably longer monitoring times), but will provide a better overall assessment than a network targeted only at maximum response sites.

3.5   ANCILLARY DATA

For each of the suggested water and sediment indicators, there are possible confounding environmental factors that are not related to changes in Hg loading, but that could result in responses in some of the indicators. Several of these confounding factors have been previously mentioned, but a general discussion of them is provided here.

Of the previously discussed indicators, those that involve MeHg (MeHg and %MeHg in sediment and water, and instantaneous methylation rate) are the most suspect to confounding effects. Mercury methylation is driven by a complex set of factors, but can be generally grouped into those that affect the activity of methylating and demethylating bacteria, and those that affect the abundance of bioavailable inorganic Hg. A detailed review of these processes can be found in Benoit et al. (2003), but are briefly explained here. Sulfate reducing bacteria (SRB) are generally thought to be the most important methylating agents in the environment; thus, the sulfate concentration in the water column should be monitored. In addition to sulfate, which is chemically reduced during sulfate reduction, SRB require a labile carbon source to serve as the electron donor. In addition, SRB respire sulfide, which has been suggested to play a direct role in regulating Hg bioaccessibility. Thus, pore water sulfide levels should be evaluated when sediment MeHg and methylation rate assays are performed. In most cases, carbon is not limiting at sites of methylation, but changes to carbon fluxes and/or composition due to disturbances such as eutrophication could affect net MeHg production. As such, nutrient status at monitoring sites should be measured. Although a definitive relation between temperature and MeHg abundance is not evident, temperature is a fundamentally important component of all environmental studies and should be monitored.

Several water chemistry factors that are thought to affect availability of Hg to methylating microbes have been shown to relate to MeHg abundance, including:

pH, DOM, alkalinity, major and trace metals, and sulfide. The details of how each of these constituents affect inorganic Hg bioaccessibility and MeHg production is not necessarily clear or predictable. For example, DOM appears to increase Hg bioaccessibility such that we might expect increasing DOM to result in greater MeHg. However, DOM also strongly limits UV light penetration, which has been shown to be an effective demethylation (conversion of MeHg to Hg) agent in aquatic ecosystem. For the purposes of a monitoring program whose purpose is to document change, we believe it is important to include all these factors in the monitoring plan.

In addition to eutrophication, other catchment disturbances — both natural (fire, drought, beaver impoundment) and anthropogenic (logging, farming, urban development) — can greatly alter the export of soil Hg to downstream lakes. Increased sediment mobilization can result in greater contributions of relic Hg to downstream water bodies, and thus the appearance of a greater Hg load, but that is not related to recent changes in atmospheric emissions. Increased sediment loads also result in high sediment mass accumulation rates, which would give the appearance of lower concentrations in recent sediments if the sediment accumulation rate is not accounted for in monitoring efforts.

3.6 ANTICIPATED RESPONSE TIMES

Monitoring programs should be designed to track HgT and MeHg concentrations and trends, and help estimate fluxes in terrestrial and aquatic systems following changes in atmospheric emissions. It will be critical to monitor for time periods that allow systematic changes in concentrations to be distinguished from year-to-year variability, and allow the system to approach a new steady-state set of concentrations in response to different loading conditions. How rapidly Hg and MeHg concentrations and methylation rates will take to respond is not well established, but we can expect that there will be a large range (a few years to many decades) in recovery periods among various ecosystem types. Furthermore, the rate of response may change with time. To illustrate this point, Figure 3.12 shows a simple exponential decline in concentration that might be predicted for a hypothetical well-mixed compartment (e.g., lake) following a one-time step decline in loading, where the removal mechanisms (e.g., outflow or burial) are directly related to the concentration (first order). Real ecosystems, however, tend to involve series of compartments that are linked in complex ways and that can include feedback mechanisms. These links can result in responses that are more complicated. If, for example, a water body received mercury from both the atmosphere and terrestrial runoff (a very common situation), and the terrestrial system imposes a time lag on the delivery of atmospheric deposition to a lake, a lake might experience a "2-phase" response to a change in atmospheric deposition. There would be an initial, more rapid component of the response due to the change in loading directly to the lake surface, and a second slower phase due to the slower change in loading from terrestrial runoff (Figure 3.12). Similarly, contaminated sediments could potentially feed mercury back to overlying waters for long periods following reductions in external mercury loads to a system. Two-phase response patterns have been observed — for example, following reductions in Hg releases from a chor-alkali facility in Ontario, Canada (Figure 3.1). These

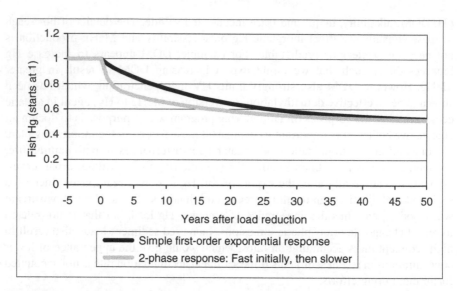

FIGURE 3.12 Simple exponential and 2-phase response dynamics model showing a response of hypothetical ecosystem to a change in Hg load.

linkages and complexities in natural systems mean that the rate of recovery observed initially following changes in atmospheric deposition may not represent the long-term response trajectory. As a result, monitoring programs should be continued long enough to accommodate complex response dynamics.

ACKNOWLEDGMENTS

This manuscript benefited substantially from the comments provided by 3 anonymous reviewers. Funding support from the US Geological Survey's Toxics and Priority Ecosystems programs to D. Krabbenhoft, and from NSF grant No. 0451345 to C. Gilmour and R. Mason are greatly appreciated.

REFERENCES

Appleby PG. 2001. Chronostratigraphic techniques in recent sediments. In: Last WM, Smol JP, editors, Tracking environmental change using lake sediments. Vol. 1: Basin analysis, coring, and chronological techniques. Dordrecht, the Netherlands: Kluwer Academic Publishers, p. 171–203.

Atkeson T, Axelrad D, Pollman C, Keeler G. 2003. Integrating atmospheric mercury deposition and aquatic cycling in the Florida Everglades: an approach for conducting a total maximum daily load analysis for an atmospherically derived pollutant. Tallahassee (FL): Florida Department of Environmental Protection (FDEP) (http://www.floridadep.org/labs/mercury/index.htm).

Babiarz CL, Hurley JP, Benoit JM, Shafer MM, Andren AW, Webb DA. 1998. Seasonal influences on partitioning and transport of total and methylmercury in rivers from contrasting watersheds. Biogeochemistry 41:237–257.

Babiarz CL, Andren AW. 1995. Total concentrations of mercury in Wisconsin (USA) lakes and rivers. Water Air Soil Pollut 83:173–183.

Babiarz CL, Hurley JP, Hoffmann SR, Andren AW, Shafer MM, Armstrong DE. 2001. Partitioning of total mercury and methylmercury to the colloidal phase in fresh waters. Environ Sci Technol 35:4773–4782.

Balogh SJ, Engstrom DR, Almendinger JE, Meyer ML, Johnson DK. 1999. A history of mercury loading in the upper Mississippi River reconstructed from the sediments of Lake Pepin. Environ Sci Technol 33:3297–3302.

Benoit JM, Fitzgerald WF, Damman AWH. 1998. The biogeochemistry of an ombrotrophic bog: evaluation of use as an archive of atmospheric mercury deposition. Environ Res 78:118–133.

Benoit J, Gilmour CC, Heyes A, Mason RP, Miller C. 2003. Geochemical and biological controls over methylmercury production and degradation in aquatic ecosystems. In: Chai Y, Braids OC, editors, Biogeochemistry of environmentally important trace elements, ACS Symposium Series #835. Washington, D.C.: American Chemical Society, p. 262–297.

Bindler R, Renberg I, Appleby PG, Anderson NJ, Rose NL. 2001. Mercury accumulation rates and spatial patterns in lake sediments from west Greenland: a coast to ice margin transect. Environ Sci Technol 35:1736–1741.

Bloom NS. 1992. On the chemical form of mercury in edible fish and marine invertebrate tissue. Can J Fish Aquat Sci 49:1010–1017.

Bloom NS, Fitzgerald WF. 1988. Determination of volatile mercury species at the picogram level by low temperature gas chromatography with cold-vapor atomic fluorescence detection. Anal Chim Acta 208:151–161.

Bloom NS, Gill GA, Cappellino S, Dobbs C, McShea L, Driscoll C, Mason R, Rudd JWM. 1999. Speciation and cycling of mercury in Lavaca Bay, Texas, sediments. Environ Sci Technol 33:7–13.

Bodaly RA, Rudd JWM, Flett RJ. 1998. Effect of urban sewage treatment on total and methylmercury concentrations in effluents. Biogeochemistry 40:279–291.

Brumbaugh WG, Krabbenhoft DP, Helsel DR, Wiener JG, Echols K. 2001. A national pilot study of mercury contamination of aquatic ecosystems along multiple gradients: bioaccumulation in fish, U.S. Geological Survey Water-Biological Science Report BSR-2001-009.

DeWild JF, Olson ML, Olund SD. 2002. Determination of methyl mercury by aqueous phase ethylation, followed by gas chromatographic separation with cold vapor atomic fluorescence detection, U.S. Geological Survey Open File Report 01-445, 23 p.

Domagalski J. 1998. Occurrence and transport of total mercury and methyl mercury in the Sacramento River Basin, California. J Geochem Explor 64:277–291.

Engstrom DR, Swain EB. 1997. Recent declines in atmospheric mercury deposition in the upper Midwest. Environ Sci Technol 312:960–967.

Engstrom DR, Pollman CD, Fitzgerald WF, Balcom PH. 2003. Evaluation of recent trends in atmospheric mercury deposition in south florida from lake-sediment records. Tallahassee (FL): Florida Department of Environmental Protection, 27 pp.

[EPRI] Electric Power Research Institute. 1996. Protocol for estimating historic atmospheric mercury deposition EPRI/TR-106768.

Fitzgerald WF, Engstrom DR, Mason RP, Nater EA. 1998. The case for atmospheric mercury contamination in remote areas. Environ Sci Technol 32:1–7.

Furutani A, Rudd JWM. 1980. Measurement of mercury methylation in lakewater and sediment samples. Appl Environ Microbiol 40:770–776.

Ganguli PM, Mason RP, Abu-Saba KE, Anderson RS, Flegal AR. 2000. Mercury speciation in drainage from the New Idria mercury mine, California. Environ Sci Technol 34:4773–4779.

Gill GA, Bruland KW. 1990. Mercury speciation in surface freshwater systems in California and other areas. Environ Sci Technol 24:1392–1400.

Gilmour CC, Riedel GS. 1995. Measurement of Hg methylation in sediments using high specific-activity 203Hg and ambient incubation. Water Air Soil Pollut 80:747–756.

Gilmour CC, Riedel GS, Ederington MC, Bell JT, Benoit JM, Gill GA, and Stordal MC. 1998. Methylmercury concentrations and production rates across a trophic gradient in the northern Everglades, Biogeochemistry 40:327–345.

Gray JE, Theodorakos PM, Bailey EA, Turner RR. 2000. Distribution, speciation, and transport of mercury in stream-sediment, stream-water, and fish collected near abandoned mercury mines in southwestern Alaska, USA. Sci Total Environ 260:21–33.

Henry EA, Dodge-Murphy LJ, Bigham GN, Klein SM, Gilmour CC. 1995. Total mercury and methylmercury mass balance in an alkaline, hypereutrophic urban lake (Onondaga Lake, N.Y.). Water Air Soil Pollut 80:489–498.

Hintelmann H, Evans RD. 1997. Application of stable isotopes in environmental tracer studies — measurement of monomethylmercury (CH_3Hg^+) by isotope dilution ICP-MS and detection of species transformation. Fresenius J Anal Chem 358:378–385.

Hintelmann H, Harris, R, Heyes A, Hurley J, Kelly C, Krabbenhoft D, Lindberg S, Rudd J, Scott K, St. Louis V. 2002. Reactivity and mobility of new and old mercury deposition in a boreal forest ecosystem during the first year of the METAALICUS study. Environ Sci Technol 36:5034–5040.

Hurley JP, Watras CJ, Bloom NS. 1994. Distribution and flux of particulate mercury in four stratified seepage lakes. In: Watras CJ, and Huckabee J, editors, Mercury as a global pollutant: integration and synthesis. Chelsea (MI): Lewis, p. 69–82.

Hurley JP, Benoit, JM, Babiarz CL, Shafer MM, Andren AW, Sulivan JR, Hammond R, Webb D. 1995. Influences of watershed characteristics on mercury levels in Wisconsin rivers. Environ Sci Technol 29:1867–1875.

Hurley JP, Cowell SE, Shafer MM, Hughes PE. 1998. Partitioning and transport of total and methylmercury in the lower Fox River, WI. Environ Sci Technol 32:1424–1432.

Hurley JP, Cowell SE, Shafer MM, Hughes PE. 1998. Tributary loading of mercury to Lake Michigan: importance of seasonal events and phase partitioning. Sci Total Environ 213:129–137.

Hydro Québec, Genivar. 1997. Summary Report. Evolution of fish mercury levels at the La Grande complex, Québec. (1978–1994).

Kamman NC, Engstrom DR. 2002. Historical and present fluxes of mercury to Vermont and New Hampshire lakes inferred from 210Pb dated sediment cores. Atmos Environ 36:1599–1609.

Kelly CA, Rudd JWM, St. Louis VL, Heyes A. 1995. Is total mercury concentration a good predictor of methylmercury concentration in aquatic systems? Water Air Soil Pollut 80:715–724.

Kelly CA, Rudd JWM, Bodaly RA, Roulet NP, St. Louis VK, Heyes A, Moore TR, Schiff S, Aravena R, Scott K, Dyck B, Harris R, Warner B, Edwards G. 1997. Increases in fluxes of greenhouse gases and methylmercury following flooding of an experimental reservoir. Environ Sci Technol 31:1334–1344.

Krabbenhoft DP, Babiarz CL. 1992. The role of groundwater transport in aquatic mercury cycling. Water Resources Res 28:3119–3128.

Krabbenhoft DP, Beniot JM, Babiarz CL, Hurley JP, Andren AW. 1995. Mercury cycling in the Allequash Creek Watershed Water Air Soil Pollut 80:425–433.

Krabbenhoft DP, Gilmour CC, Beniot JM, Babiarz CL, Andren AW, Hurley JP. 1998a. Methylmercury dynamics in littoral sediments of a temperate seepage lake. C J Fish Aquat Sci 55:835–844.

Krabbenhoft DP, Hurley JP, Olson ML, Cleckner LB. 1998b. Diel variability of mercury phase and species distributions in the Florida Everglades. Biogeochemistry 40:311–325.

Krabbenhoft DP, Wiener JG, Brumbaugh WG, Olson ML, DeWild JF, Sabin T. 1999. In: Morganwalp DW, Buxton HT, editors, A national pilot study of mercury contamination of aquatic ecosystems along multiple gradients. U.S. Geological Survey Toxic Substances Hydrology Program — Proceedings of the Technical Meeting, Charleston, SC, March 8–12, 1999. Vol 2: Contamination of hydrologic systems and related ecosystems, U.S. Geological Survey Water-Resources Investigations Report 99-4018B, p. 147–160.

Krabbenhoft DP, Orem W, Aiken G, Gilmour CG. 2004. Unraveling the complexities of mercury methylation in the Everglades: the use of mesocosms to test the effects of "new" mercury, sulfate, and organic carbon. Proc 7th Int Conf Mercury Pollut, RMZ-MG, 51(1–3):June 2004.

Lamborg CH, Fitzgerald WF, Damman AWH, Benoit JM, Balcom PH, Engstrom DR. 2002. Modern and historic atmospheric mercury fluxes in both hemispheres: global and regional mercury cycling implications. Global Biogeochem Cycles 16:1104; doi:1110.1029/2001GB1847.

Lee YH, Bishop K, Petterson C, Iverfeldt A, Allard B. 1995. Subcatchment output of mercury and methylmercury at Svartberget in Northern Sweden. Water Air Soil Pollut 80:455–465.

Long ED, MacDonald D, Smith SL, Calder FD. 1995. Incidence of adverse biological effects within ranges of chemical concentrations in marine and estuarine sediments. Environ Manage 19:81–97.

Mason RP, Rolfhus KR, Fitzgerald WF. 1998. Mercury in the North Atlantic. Mar Chem 61:37–53.

Mierle G, Ingram R. 1991. The role of humic substances in the mobilization of mercury from watersheds. Water Air Soil Pollut 56:349–357.

Munthe J, Hultberg H. 2004. Mercury and methylmercury in runoff from a forested catchment — concentrations, fluxes, and their response to manipulations. Water, Air Soil Pollut: Focus 4:607–618.

Olson ML, DeWild JF. 1999. Low-level techniques for the collection and species-specific analysis of low levels of mercury in water, sediment, and biota. In: Morganwalp, DW, Buxton HT, editors, U.S. Geological Survey Toxic Substances Hydrology Program — Proceedings of the Technical Meeting, Charleston, SC, March 8–12, 1999, Vol 2 of 3: Contamination of hydrologic systems and related ecosystems: U.S. Geological Survey Water-Resources Investigations Report 99-4018B, p. 191–200.

Olund SD, DeWild JF, Olson ML, Tate MT. 2004. Methods for the preparation and analysis of solids and suspended solids for total mercury, U.S. Geological Survey Techniques and Methods Report 5A-8, 23 p.

Parks J, Hamilton A. 1987. Accelerating the recovery of the mercury contaminated Wabigoon English River system. Hydrobiologia 149:159–188.

Ravichandran M, Aiken GR, Reddy MM, Ryan JN. 1998. Enhanced dissolution of cinnabar (mercuric sulfide) by organic matter from the Florida Everglades. Environ Sci Technol 32:3305–3311.

Rolfhus KR, Sakamoto HE, Cleckner LB, Stoor RW, Babiarz CL, Back RC, Manolopoulos H, Hurley JP. 2003. The distribution of mercury in Lake Superior. Environ Sci Technol 37:865–872.

Rudd JWM, Turner MA, Furutani A, Swick AL, Townsend BE. 1983. The English-Wabigoon River system. I. A synthesis of recent research with a view towards mercury amelioration. Can J Fish Aquat Sci 40:2206–2217.

Schuster PF, Krabbenhoft DP, Naftz DL, Cecil LD, Olson ML, DeWild JF, Susong DD, Green JR, Abbott ML. 2002. Atmospheric mercury deposition during the last 270 years: a glacial ice core record of natural and anthropogenic sources. Environ Sci Technol 36:2303–2310.

St. Louis VL, Rudd JWM, Kelly CA, Beaty KG, Bloom NS, and Flett RJ. 1994. Importance of wetlands as sources of methyl mercury to boreal forest ecosystems, Can J Fish Aquat Sci 51:1065–1076.

St. Louis VL, Rudd JWM, Kelly CA, Beaty KG, Flett RJ, Roulet NT. 1996. Production and loss of methylmercury and loss of total mercury from boreal forest catchments containing different types of wetlands. Environ Sci Technol 30:2719–2729.

St. Louis VL, Rudd JWM, Kelly CA, Bodaly RA, Paterson MJ, Beaty KG, Hesslein RH, Heyes A, Majewski AR. 2004. The rise and fall of mercury methylation in an experimental reservoir. Environ Sci Technol 38:1348–1358.

Swain EB, Engstrom DR, Brigham ME, Henning TA, Brezonik PL. 1992. Increasing rates of atmospheric mercury deposition in midcontinental North America. Science 257:784–787.

[USEPA] US Environmental Protection Agency. 1996. Method 1631: mercury in water by oxidation, purge and trap, and cold vapor atomic fluorescence (CVAFS). Draft method EPA 821-R-96-012.

Waldron MC, Colman JA, and Breault RF. 2000. Distribution, hydrologic transport, and cycling of total mercury and methyl mercury in a contaminated river-reservoir-wetland system (Sudbury River, eastern Massachusetts), Can J Fish Aquat Sci 57:1080–1091.

Wang W, Driscoll CT. 1995. Patterns of total mercury concentrations in Onondaga Lake, New York. Environ Sci Technol 29:2261–2266.

Watras CJ, Bloom NS, Hudson RJM, Gherini S, Munson R, Claas SA, Morrison KA, Hurley JP, Wiener JG, Fitzgerald WF, Mason RP, Vandal G, Powell D, Rada R, Rislove L, Winfrey M, Elder J, Krabbenhoft DP, Andren AW, Babiarz C, Porcella DB, Huckabee JW. 1994. Sources and fates of mercury and methylmercury in Wisconsin lakes. In: Watras CJ, Huckabee J, editors, Mercury as a global pollutant: integration and synthesis. Chelsea (MI): Lewis, p. 153–177.

Watras CJ, Morrison KA, Hudson RJM, Frost TM, Kratz TK. 2000. Decreasing mercury in northern Wisconsin: temporal patterns in bulk precipitation and a precipitation-dominated lake. Environ Sci Technol 34:4051–4057.

Wiener JG, Spry DJ. 1996. Toxicological significance of mercury in freshwater fish. In: Beyer WN, Heinz GH, Redmon-Norwood AW, editors, Environmental contaminants in wildlife: interpreting tissue concentrations. Boca Raton (FL): Lewis Publishers, p. 297–339.

Wiener JG, Krabbenhoft DP, Heinz GH, Scheuhammer AM. 2003. Ecotoxicology of mercury. In: Hoffman DJ, Rattner BA, Burton GA Jr, Cairns J Jr, editors, Handbook of ecotoxicology, 2nd ed. Boca Raton (FL): CRC Press, p. 409–463.

4 Monitoring and Evaluating Trends in Methylmercury Accumulation in Aquatic Biota

James G. Wiener, R.A. Bodaly, Steven S. Brown, Marc Lucotte, Michael C. Newman, Donald B. Porcella, Robin J. Reash, and Edward B. Swain

ABSTRACT

The monitoring of mercury in aquatic food webs supporting the production of fish and wildlife is directly relevant to concerns about health and ecological risks of methylmercury (MeHg) exposure. We present a framework for monitoring concentrations of mercury in aquatic biota, with emphasis on assessing responses to changes in loadings of mercury from atmospheric deposition and other sources. In this chapter, we (1) identify specific attributes (criteria) of indicators that would be useful for discerning temporal trends and spatial patterns in the concentration of mercury in aquatic biota, (2) critically evaluate and rank candidate biological indicators useful for monitoring trends in mercury, (3) outline approaches for sampling and analysis of recommended biological indicators, (4) identify ancillary data needs and potential confounding factors that should be considered or documented to ensure the defensible interpretation of data on monitored biological indicators, and (5) consider the environmental settings (waterbody type and geographic location) that would be most sensitive for detecting changes in atmospheric deposition of mercury. Criteria were applied to ensure that the biological indicators selected are useful, relevant, and sufficiently diagnostic to detect a change in mercury bioaccumulation in response to altered mercury loadings. Toxicological problems with mercury in aquatic ecosystems result from biotic exposure to MeHg, a highly toxic compound that readily accumulates in exposed organisms and can biomagnify to high concentrations in organisms atop aquatic food webs. Biotic monitoring should, therefore, focus on assessing trends in bioaccumulation of MeHg; in samples from trophic levels below fish, this requires the determination of MeHg. We considered six general groups of aquatic biological indicators: piscivorous fish, prey fish, benthic invertebrates, zooplankton, phytoplankton, and periphyton. Piscivorous fish and 1-year-old prey fish, all analyzed individually, are considered the preferred aquatic biological indicators for trend monitoring. For piscivorous fish, total-mercury determinations on axial muscle (preferably without skin), sampled annually, would indicate gradual (multiyear) trends in MeHg that are directly relevant to humans who eat sport fish. For prey fish, annual sampling and analysis of either whole fish or axial muscle for total mercury should indicate annual changes in exposure to MeHg. In North America, the historical record for MeHg in piscivorous fish (from data on total mercury in filets or axial muscle) extends about 35 years, much longer than the comparatively sparse historical record for MeHg in water and aquatic biota of lower trophic levels. The analytical method (cold vapor atomic absorption spectrophotometry) that produced most of the historic data on total mercury in piscivorous fish is valid, and the potential utility of these existing data for trend analysis merits careful consideration. Benthic invertebrates have been monitored and analyzed for mercury more extensively in estuarine systems than in fresh waters. The consumption of estuarine macroinvertebrates, such as oysters, clams, shrimp, and crabs, is also a direct pathway for human exposure to MeHg. Determination of MeHg in shellfish and macrocrustaceans could, therefore, be useful for trend monitoring in estuaries. The importance of MeHg uptake and transfer at the base of the food web is recognized. However, the utility of periphyton, phytoplankton, zooplankton, and freshwater benthic invertebrates for trend monitoring is diminished by interpretational complexities associated with large temporal variation in the biotic composition and MeHg content among samples, and by our limited understanding of the processes and variables that affect concentrations of MeHg in these groups. Many anthropogenic and natural factors, independent of the bulk loading of mercury from atmospheric deposition, can strongly influence the concentrations of MeHg in aquatic biota. Our ability to discern linkages between MeHg concentrations in aquatic biota and changing external loadings of mercury to aquatic systems will depend on knowledge of such factors and on the minimization of their confounding effects in biotic monitoring programs.

4.1 INTRODUCTION

Elevated mercury concentrations in fish tissue adversely affect the quality of fishery resources in many inland, coastal, and marine waters. In the United States, mercury was responsible for 76% of all fish-consumption advisories in 2004, and 44 states and 1 territory had advisories attributed to mercury (USEPA 2005). Growing awareness of the mercury problem has prompted increasing efforts to survey mercury in fish, and the number of statewide fish-consumption advisories issued for coastal waters, lakes, and rivers in the United States has increased substantially during the past decade (Wiener et al. 2003; USEPA 2005). In Canada, mercury accounted for more than 97% (2572) of all fish-consumption advisories listed in 1997 (USEPA 2001). Nearly all of the mercury in fish is methylmercury (MeHg), a highly toxic compound that readily crosses biological membranes, accumulates in exposed organisms, and can biomagnify to high concentrations in aquatic food webs (Grieb et al. 1990; Bloom 1992; Francesconi and Lenanton 1992; Wiener et al. 2003).

Atmospheric deposition is an important source of total mercury in many surface waters (Fitzgerald et al. 1998; Bindler et al. 2001; Lin et al. 2001; Lamborg et al. 2002), and it has been widely inferred that a significant portion of the MeHg bioaccumulated in many aquatic ecosystems is derived from mercury entering the surface water or its watershed in atmospheric deposition (Johansson et al. 1991; Watras et al. 1994; Rolfhus and Fitzgerald 1995; Jackson 1997; Monteiro and Furness 1997; Downs et al. 1998). In many remote and semi-remote areas of the Northern Hemisphere that lack in-watershed sources of anthropogenic mercury, the rate of mercury accumulation in lacustrine sediments has increased by a factor of 2 to 5 or more since the mid-1800s or early 1900s, based on analyses of dated cores of sediment and peat (Swain et al. 1992; Lucotte et al. 1995; Lockhart et al. 1998; Lorey and Driscoll 1999; Bindler et al. 2001; Lamborg et al. 2002; Shotyk et al. 2005). Some cores from semi-remote sites show evidence of recent declines in mercury deposition, possibly associated with decreasing emissions of anthropogenic mercury (Engstrom and Swain 1997; Benoit et al. 1998; Bindler et al. 2001; Shotyk et al. 2003, 2005).

We present a framework for monitoring concentrations of mercury in aquatic biota, with emphasis on assessing responses to changes in loadings of mercury from atmospheric deposition and other sources. The monitoring of mercury in aquatic food webs supporting the production of fish and wildlife is directly relevant to societal concerns about this toxic metal. Much of the scientific effort on mercury contamination of aquatic food webs has been prompted by the human health risks of MeHg exposure (Myers and Davidson 1998; Mahaffey 2000; Clarkson 2002), given that the consumption of finfish and shellfish is the primary exposure pathway in humans (NRC 2000; Mahaffey et al. 2004). Consumption of fish is also an important pathway of MeHg exposure for wildlife atop aquatic food webs (Heinz 1996; Wolfe et al. 1998; Wiener et al. 2003).

4.2 OBJECTIVES

This chapter focuses on monitoring trends in bioaccumulation in relation to anticipated changes in emissions of mercury from anthropogenic sources. Aquatic biota, however, are exposed to mercury from multiple sources, including historic anthropogenic, current anthropogenic, and natural sources. We identify aquatic biological indicators that can provide evidence of a temporal change in bioaccumulation of mercury (estimated from concentrations in tissue or whole organisms) from all sources. The objectives of this chapter are fivefold:

1) To identify specific attributes (criteria) of indicators that would be useful for discerning temporal trends and spatial patterns in the concentration of mercury in aquatic biota
2) To critically evaluate and rank candidate biological indicators useful for monitoring trends in mercury
3) To outline approaches for sampling and analysis of recommended biological indicators

4) To identify ancillary data needs and potential confounding factors that should be considered or documented to ensure the defensible interpretation of data on monitored biological indicators

5) To identify the environmental settings (water body type and geographic location) that would be most sensitive for detecting changes in atmospheric deposition of mercury in a trend-monitoring program

Many factors other than the bulk loading of mercury from atmospheric deposition can strongly influence the concentrations of MeHg in aquatic biota. Our ability to discern trends in MeHg concentrations in aquatic biota that are linked to changing loadings of mercury will depend on knowledge of such factors and on the minimization of their confounding effects in biotic monitoring programs for mercury.

4.3 AQUATIC BIOLOGICAL INDICATORS

4.3.1 CRITERIA TO SELECT INDICATORS

The selection of biological indicators should be guided by criteria to ensure that the indicators are relevant, useful, and sufficiently diagnostic to detect a change in the concentration of mercury in whole organisms or specific tissue(s) over multi-year or decadal time scales. We identified 9 criteria for the selection of biological indicators:

1) *Relevance.* A key criterion in the selection of biological indicators is relevance to human and ecological health and to the development of policy. Fish are directly relevant, for example, given that consumption of fish is the primary pathway for exposure to MeHg. The concentration of MeHg in fish is also a key variable in the issuance of fish-consumption advisories.

2) *Historical data on the indicator.* Existing information on the statistical variation, bias, and other interpretational attributes of potential biological indicators should be examined and considered in the design of a sampling program for assessing trends in mercury bioaccumulation.

3) *Clear recognition of confounding factors.* Many human and natural factors that are unrelated to bulk loadings of mercury from the atmosphere or other sources can strongly influence concentrations of mercury in aquatic biota. A *confounding factor* is here defined as a variable that interferes with the isolation of the effects of mercury loading on temporal trends in mercury concentrations in monitored biota within a water body or group of waters. Our ability to discern biotic trends in mercury concentrations linked to changing loadings of mercury to aquatic systems will require both knowledge of potential confounding factors and the minimization of their confounding effects in monitoring programs (see Section 4.6.2).

4) *Knowledge of intrinsic co-variables.* Concentrations of mercury in fish are typically correlated with age or body size. An understanding of, and ability to account for, the effects of such intrinsic variables is essential for evaluating contaminant trends.

5) *Broad geographic distribution.* When feasible, monitoring should focus on bioindicator species or taxa having a broad geographic range, to assess trends and patterns across large geographic areas.

6) *Importance in trophic transfer of MeHg.* Preference should be given to bioindicator organisms that have an important role in the trophic transfer of MeHg within food webs.

7) *Constrained feeding ecology and trophic position.* Temporal shifts in feeding habits or trophic position can alter dietary MeHg uptake, complicating the interpretation of trends in mercury concentration. An *a priori* knowledge of feeding ecology should facilitate the selection of suitable candidate bioindicator species during the design of a monitoring program. If a bioindicator is known to undergo ontogenetic shifts in diet, it would be advisable to limit sampling and analysis to a given life stage.

8) *Temporal response to changes in mercury loadings.* Long-lived bioindicators with long response times (several years) to changing mercury loadings could be more susceptible to the influence of confounding factors, possibly reducing their utility for detecting effects of altered loadings. Thus, short-lived ($\leq$1-year old), as well as long-lived, organisms should be considered when selecting potential bioindicators.

9) *Impacts of sampling on the target population.* Continued removal of individuals could eventually affect the concentration of mercury in members of the sampled population, creating artifacts in trend data. The potential effects of sampling on the target population, and potential approaches for reducing such effects, should be considered. Nonlethal techniques for sampling of fish tissue (Baker et al. 2004; Peterson et al. 2005) may be required when sampling protected populations or when sampling in protected environments, such as national parks.

4.3.2 CANDIDATE AQUATIC BIOLOGICAL INDICATORS

Six general groups of aquatic biological indicators were considered: 1) piscivorous fish, 2) prey fish, 3) benthic invertebrates, 4) zooplankton, 5) phytoplankton, and 6) periphyton. We begin with a brief summary of our understanding of mercury bioaccumulation and trophic transfer in aquatic ecosystems, to provide essential background information for the subsequent sections on aquatic biological indicators. This summary is based on selected recent reviews (Jackson 1998; Morel et al. 1998; Benoit et al. 2003; Wiener et al. 2003) and the original reports cited.

In a toxicological sense, the primary problem with mercury in aquatic ecosystems can be defined as biotic exposure to, or bioaccumulation of, MeHg. Trend monitoring of mercury in aquatic biota should accordingly focus on MeHg, which readily crosses biological membranes and accumulates to concentrations in aquatic organisms that vastly exceed those in surface water. In fish, concentrations of MeHg commonly exceed those in water by a factor of 10^6 to 10^7 or more. Most of the mercury in the aquatic environment is inorganic, yet nearly all (>95%) of the mercury accumulated in fish is MeHg (Grieb et al. 1990; Bloom 1992; Francesconi and Lenanton 1992), obtained almost entirely via dietary uptake (Rodgers 1994; Hall et al. 1997; Harris and Bodaly 1998). Relative to MeHg, inorganic mercury is less readily transferred through successive trophic levels and does not biomagnify (Watras et al. 1998).

Methylmercury biomagnifies in aquatic food webs, and patterns in biomagnification are similar even among aquatic systems that differ in type of water body, mercury source, and pollution intensity. The transfer of MeHg in the upper trophic levels of aquatic food webs is almost entirely via dietary uptake, whereas direct uptake from water can be important for some lower food-chain organisms, such as phytoplankton and zooplankton. The concentration of MeHg increases up the food web from water and lower trophic levels to fish and piscivores, and the fraction of total mercury present as MeHg also increases with increasing trophic level through fish. The fraction of total mercury present as MeHg can vary greatly within trophic levels below fish. In aquatic invertebrates, for example, the MeHg fraction can range from about 10% to more than 90% of total mercury. It is, therefore, essential to determine MeHg (rather than total mercury) in biological samples from trophic levels below fish, including phytoplankton, periphyton, benthic invertebrates, and zooplankton. In fish, determination of total mercury, which requires less analytical effort and expense than MeHg, provides reliable estimates of MeHg concentration.

The greatest increase in MeHg concentration occurs in the trophic step between water and algae. Bioaccumulation factors between water and seston, for example, often range from about 10^5 to about 10^6, whereas ratios of MeHg concentrations between successive trophic levels above algae are generally less than 10^1. Within an assemblage of fish, concentrations of MeHg increase with ascending trophic level, and variation in trophic position accounts for much of the variation in mercury concentration among species within a given water body. Concentrations of MeHg in fish also increase with increasing age or size because of the very slow rate of elimination relative to the rapid rate of uptake and because larger fish can consume larger prey with higher concentrations of MeHg. Much of the MeHg accumulated in fish is stored in skeletal muscle, tightly bound to sulfhydryl groups in protein.

Although the entry of MeHg into the base of the food web and its subsequent transfer in the lowest trophic levels are poorly understood, it is evident that the concentration of MeHg in all trophic levels is strongly correlated with its supply from methylating environments. In fish, for example, much of the modern spatial variation in mercury concentrations (within a given trophic level) can be attributed to variation in factors and processes that affect the microbial production of MeHg and its entry into oxic waters.

4.3.2.1 Fish

We consider 2 groups of fish as candidate biological indicators for monitoring trends in MeHg: 1) piscivorous (fish-eating) fish and 2) small prey fish (often termed "forage fish"). Analyses of piscivorous fish reflect the concentrations of MeHg near the top of aquatic food webs, providing a useful measure of potential dietary exposure in humans who eat fish. Small prey fish are eaten by a variety of piscivorous fish, birds, and mammals, but are generally not important in the diet of humans. Humans eat large omnivorous fish; however, the substantial confounding effects of the large spatial and temporal variation in the diet and trophic position of omnivorous fish diminish their potential utility for trend analysis.

4.3.2.1.1 Piscivorous Fish

Piscivorous fish can accumulate high concentrations of MeHg, and information on MeHg in piscivorous fish is directly relevant to the public and the policy community. Piscivorous fish are present in most surface waters and can be obtained with moderate sampling effort with a variety of active and passive gear. Sampling would generally not affect target populations except in very small lakes and streams, where nonlethal sampling would lessen impacts on target populations.

In the United States, the threshold mercury concentration for commercial sale of fish is determined by the Food and Drug Administration, whereas consumption advice for recreational (noncommercial) fish is developed by individual states and tribes. Mercury data collected for development of fish-consumption advisories are typically from analyses of filets (axial muscle tissue, with or without skin) for total mercury, with concentrations expressed on a wet-weight basis. Analysis of filets for total mercury yields a valid estimate of MeHg concentration (Grieb et al. 1990; Bloom 1992), whether the analyzed sample consists of a large filet or a small mass of tissue obtained with a biopsy needle (Cizdziel et al. 2002; Baker et al. 2004).

Many piscivorous fish are important recreationally or commercially. The sampling and analysis of heavily exploited fish stocks are not recommended for trend monitoring, because intensive fishing pressure — over a period of years to decades — can substantially reduce mercury concentrations in members of heavily exploited populations (Section 4.6.1). Interpretation of temporal trends in mercury concentrations in fish, relative to changes in mercury loadings, will be more defensible if applied to fish populations and water bodies that are not subjected to intensive fishing pressure. Many commercially and recreationally important marine fish have been significantly depleted by over-fishing (Pauly et al. 2003; Coleman et al. 2004; Hutchings and Reynolds 2004). The application of trend data on mercury concentrations in heavily exploited marine fishes, such as yellowfin tuna *Thunnus albacares* (Kraepiel et al. 2003), has questionable validity as an approach for assessing temporal changes in the abundance of MeHg in the ecosystem.

In summary, piscivorous fish are present in most surface waters, require moderate sampling effort, and are the primary pathway for dietary exposure of humans to MeHg. Analyses of filets or axial muscle for total mercury should indicate gradual (multi-year) trends in the supply of MeHg. Given these attributes, nonmigratory piscivorous fish are priority candidates for monitoring MeHg. In very small lakes or streams, nonintrusive sampling methods should be used to reduce impacts on sampled populations.

4.3.2.1.2 Prey Fish

Prey fish are here defined as small, usually short-lived, finfish. In North American fresh waters, many are members of the cyprinid (minnow), percid (perch), and centrarchid (sunfish) families. Prey fish are widely distributed, common, and important in the transfer of MeHg to higher trophic levels, such as piscivorous fish and many fish-eating birds. MeHg concentrations in prey fish of uniform age are less susceptible to certain, potential confounding factors, such as variation in trophic position, than are concentrations in long-lived, piscivorous fish.

Mercury concentrations in prey fish are useful indicators of relative MeHg levels in food webs supporting the production of sport fish and wildlife, information relevant to the public and the policy community. There is a sizable scientific literature on MeHg in prey fish, but they have been monitored less extensively than sport fish. Effects of removal sampling on target populations would be insignificant in all but the very smallest lakes.

This group of fish includes species that feed on zooplankton, benthic invertebrates, and occasionally, periphyton (Roseman et al. 1996; Bodaly and Fudge 1999; Gorski et al. 1999). The size of young prey fish can vary seasonally because measurable growth occurs throughout much of the year. Small prey fish, such as yearling yellow perch (*Perca flavescens*), finescale dace (*Phoxinus neogaeus*), or mimic shiner (*Notropis volucellus*), typically grow rapidly during the summer in temperate lakes, increasing their biomass by 2- to 5-fold. Mercury concentrations are usually increasing or stable during this period of growth; consequently, the total body burden of mercury in individual fish increases substantially during summer (Bodaly and Fudge 1999; Gorski et al. 1999). Concentrations of MeHg in prey fish can be expected to vary seasonally (Figure 4.1), and such variation should be considered when crafting sampling protocols for trend monitoring. Prey fish require little to moderate sampling effort, and samples taken in the early spring or fall, when growth is slow and temporal variation in mercury concentration is less, may provide the best comparisons among years and surface waters. Analysis of prey fish for total mercury or MeHg would reveal inter-annual variation in MeHg exposure (Frost et al. 1999; Gorski et al. 1999).

In summary, prey fish are present in most surface waters, require moderate sampling effort, are important in the trophic transfer of MeHg in aquatic food webs, and probably indicate annual changes in exposure to MeHg. Given these attributes,

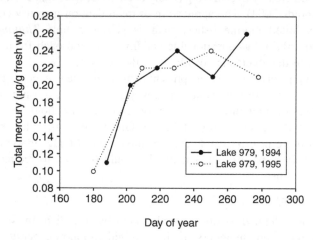

FIGURE 4.1 Whole-body concentrations of total mercury (present largely as MeHg) in caged finescale dace, showing seasonal increases in mercury concentrations during summer in Lake 979, an experimental reservoir in northwestern Ontario that was flooded during the Experimental Lakes Area Reservoir Project. (*Source:* Modified from Bodaly and Fudge 1999.)

prey fish are appropriate candidates for monitoring of MeHg. Seasonal variation should be considered carefully in the design of trend-monitoring protocols for prey fish.

4.3.2.1.3   Example of a Prey-Fish Indicator: Yellow Perch

Analyses of total mercury in whole bodies or axial muscle tissue of age-1 yellow perch have provided a useful measure of MeHg concentrations in food webs of many North American lakes. This widely distributed species inhabits lakes and reservoirs across much of the north-central, northeastern, and eastern United States and across the central and eastern provinces of Canada (Scott and Crossman 1973; Becker 1983). An ecologically similar congeneric species, the Eurasian perch (*Perca fluviatilis*), is distributed across much of Europe and northern Asia (Thorpe 1977).

During their first year, yellow perch have a small gape (jaw opening), which limits their diet largely to small zooplankton and small zoobenthos (Roseman et al. 1996; Lyons et al. 2000). Thus, the trophic position of age-1 yellow perch is not expected to vary substantially among sites. Generally abundant in lakes within much of its geographic range, the yellow perch is a preferred prey of certain piscivores, such as walleye (*Sander vitreus*) and common loons (*Gavia immer*), and is an important link in the food-web transfer of MeHg (Colby et al. 1979; Barr 1996). Concentrations of total mercury in age-1 or age-2 yellow perch are strongly and positively correlated with concentrations in coexisting piscivorous fish, including walleye, black bass (*Micropterus* spp.), and northern pike (*Esox lucius*) (Suns et al. 1987; Cope et al. 1990; JG Wiener, unpublished data for northern pike). Statistical analyses have shown strong relations between the total mercury concentration and burden of age-1 yellow perch and ecosystem characteristics (e.g., lake chemistry, wetland influence) or whole-lake manipulations (e.g., experimental acidification) that are known to influence the production of MeHg and its abundance in aquatic food webs (Grieb et al. 1990; Suns and Hitchin 1990; Wiener et al. 1990; Simonin et al. 1994; Frost et al. 1999; Wiener et al. 2003). Substantial mercury data are also available for the Eurasian perch (Metsaelae and Rask 1989; Andersson et al. 1995; Haines et al. 1995; Porvari 1998; Svobodova et al. 1999; Lindestroem 2001).

One-year-old yellow perch can be readily sampled in spring with small-mesh trap nets, seines, or small electroshockers fished in littoral habitat without significantly affecting their abundance or year-class strength. Age-1 fish obtained in spring have resided in a sampled lake for about 1 year. A target sample size of 15 to 30 whole, age-1 yellow perch (analyzed individually) from a given lake typically yields a standard error of the mean in the range of 1 to 6 ng/g wet weight, providing a precise estimate of mean whole-body concentration (JG Wiener, University of Wisconsin–La Crosse, data from Wisconsin and Minnesota lakes; corresponding mean concentrations range from about 20 to 200 ng/g wet weight). At age-1, the age of yellow perch can be accurately determined by examining scales taken near the area of insertion of the left pectoral fin.

4.3.2.2   Benthic Invertebrates

Benthic invertebrates are macroscopic animals that live at or near the sediment/water interface. Some benthic invertebrates, particularly mussels, readily accumulate metals, prompting their use as biological indicators of mercury contamination (Smith and Green 1975). Benthic invertebrates have been widely used in biological monitoring of freshwater and marine habitats (Resh and McElvary 1993; Southerland and Stribling 1995; Weigel et al. 2003), providing a foundation for their use in trend-monitoring programs.

The dietary importance of benthic invertebrates to many species of fish, birds, and mammals (Vander Zanden and Vadeboncoeur 2002) signifies their importance in the trophic transfer of MeHg and their potential relevance as biological indicators. Some benthic invertebrates (e.g., oysters, clams, shrimp, crabs, and crayfish) are consumed by humans, providing a direct pathway for exposure to MeHg. In the United States, shellfish rank below fish as a source of dietary MeHg in the human population (NRC 2000; Schober et al. 2003).

Many benthic invertebrates have short life spans (≤ 1 year) but little is known about how quickly mercury concentrations in such organisms respond to changes in *external* loadings of mercury to the aquatic ecosystem. Bed sediment can be an important sink for mercury in aquatic systems if sediment-associated mercury is isolated from active biogeochemical cycling (Henry et al. 1995; Wiener and Shields 2000). Sediment can also serve as a source of MeHg in freshwater and estuarine ecosystems, given that the oxic/anoxic interface in the sediment is an important zone of mercury methylation (Gilmour et al. 1998; Benoit et al. 2003). Benthic organisms have physical contact with bed sediment, and the relative contributions of mercury from in-place sedimentary sources and of mercury from current external sources to their MeHg burdens will influence their sensitivity as indicators to altered mercury loadings from external sources. In this regard, sediment-dwelling invertebrates that feed on particles from the overlying water column — a group including many clams and aquatic insects — may be more useful than deposit feeders as indicators of external mercury loadings. The kinetics of MeHg bioaccumulation (ingestion, assimilation, and elimination) in benthic invertebrates have received little study but the limited available information should nonetheless be applied to the selection of candidate bioindicator species and to the interpretation of trend data.

The trophic position of benthic invertebrates varies widely among species. The diet is their primary pathway of contaminant exposure, and the feeding ecology of a benthic species largely determines its exposure to dissolved and particulate sedimentary contaminants (Brown et al. 2000). The dietary assimilation efficiency for MeHg (55–70%) in marine invertebrates is much higher than that for inorganic mercury (2–22%; Wang and Fisher 1999). In marine bivalves, exposure can occur through multiple pathways, including direct uptake from the overlying water or pore water and dietary uptake via ingestion of plankton and detritus from water or sediment (Thomann et al. 1995).

Benthic invertebrate communities are taxonomically and trophically complex, and their abundance and species composition in a water body often vary seasonally and among years. Sediment-dwelling invertebrates can be readily sampled but considerable effort is often required to remove benthic organisms from grab samples of sediment, to determine their taxonomic composition, and to obtain sufficient sample mass of a target taxon for analysis. Sampling would not substantially affect target populations.

In summary, benthic invertebrates are important in the trophic transfer of MeHg. The sensitivity of benthic organisms to altered mercury loadings from external sources will presumably depend on the relative contributions of mercury from in-place sediment and external sources to their total MeHg uptake. Moreover, the large effort required to produce samples that are well-defined taxonomically and trophically reduces their desirability as biotic indicators for trend monitoring in freshwater systems. Conversely, in estuarine systems, there has been more extensive monitoring and analysis of mercury in shellfish and macrocrustaceans, which could be useful in monitoring of estuaries.

4.3.2.3 Zooplankton

Zooplankton are small, often microscopic crustaceans that live in the water column. They are widely distributed, common, and important in pelagic food webs. Zooplankton are eaten by many fish and by early life stages of some fish that become piscivorous as juveniles or adults. Zooplankton are not eaten by humans but are an appropriate and relevant candidate indicator because of their importance in the trophic transfer of MeHg to fish. Sampling would not significantly affect populations or assemblages of zooplankton, even in small lakes.

Zooplankton vary seasonally and annually in abundance (Rusak et al. 2002), and respond within hours or days to changes in the supply of MeHg to the water column (Herrin et al. 1998; Paterson et al. 1998; Tsui and Wang 2004). Zooplankton can accumulate significant quantities of MeHg from food or water (Monson and Brezonik 1999; Peech Cherewyk 2002; Tsui and Wang 2004). Spatio-temporal variation in the taxonomic composition and abundance of zooplankton is large in temperate lakes (Rusak et al. 2002). Consequently, the trophic position of bulk zooplankton samples can vary greatly because these assemblages are composed of trophically diverse taxa that can eat phytoplankton, bacteria, detritus, and other zooplankton. Species assemblages can change rapidly, and a target species or taxon may not be available at certain times of the year.

The MeHg content of zooplankton varies among taxa (Back and Watras 1995), a complicating factor that can be eliminated by the determination of MeHg in individual taxa (Back et al. 1995). Zooplankton are readily sampled but samples should be checked and processed to remove phytoplankton and detritus. A single bulk sample from a plankton net can be used to characterize the open-water community of zooplankton in a lake at a particular time. The processing of samples can require substantial effort, and bulk samples of zooplankton from some surface waters contain particles that are extremely difficult to remove, diminishing the integrity of the sample.

Existing data on MeHg in zooplankton are few, and there has been little long-term monitoring of MeHg in zooplankton. In northern lakes, the concentrations of MeHg in zooplankton vary seasonally and annually (Figure 4.2), typically increasing in summer and declining in autumn (Herrin et al. 1998; Paterson et al. 1998). In thermally stratified lakes, the concentration of MeHg in zooplankton can increase markedly after the fall overturn (Figure 4.2), when MeHg in anoxic hypolimnetic

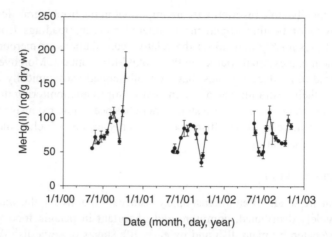

FIGURE 4.2 Concentrations of MeHg (mean ± 1 standard error) in zooplankton from Lake 240 of the Experimental Lakes Area (northwestern Ontario, Canada), showing seasonal variation during summer and pronounced rapid increases in mean concentration after the fall overturn. (*Source:* Michael J. Paterson, Fisheries and Oceans Canada, Winnipeg, Manitoba, unpublished data.)

waters becomes mixed throughout the water column (Herrin et al. 1998). Such seasonal variation merits careful attention in sampling protocols for zooplankton in a trend program for mercury, regardless of whether bulk samples or specific taxa are used as a bioindicator, and may require multiple sampling events during the year. Samples taken in mid-summer would reflect conditions for bioaccumulation of MeHg during the period of maximal growth of fish but would miss the effect of the fall overturn in stratified lakes. Given such large intra- and inter-annual variation, we expect that many years of sampling and analysis would be needed to discern long-term trends in MeHg concentrations in zooplankton.

In summary, zooplankton are present in all lakes, are readily sampled, are important in the trophic transfer of MeHg, and would respond rapidly to changes in the supply of MeHg to the water column. Seasonal and interannual changes in MeHg concentration would complicate the timing of sampling and the interpretation of trend data. Pronounced short-term variability may reduce their suitability for assessing trends on the multi-year time scales expected for reductions in mercury emissions.

4.3.2.4 Phytoplankton

Phytoplankton are microscopic plants (algae) that live in the water column and are trophically situated at or near the base of pelagic food webs. Phytoplankton are taxonomically diverse, and their abundance and species composition vary seasonally, annually, and spatially. Phytoplankton are present in all aquatic systems and are considered important in the trophic transfer of MeHg. It has been generally assumed that zooplankton receive most of their MeHg via consumption of phytoplankton; however, recent research suggests that direct uptake of MeHg by zooplankton from water may be more important than previously thought (Monson and Brezonik 1999;

Peech Cherewyk 2002; Tsui and Wang 2004). Phytoplankton have a constrained trophic position; are not consumed by fish, wildlife, or humans, and are indirectly relevant to the public or the policy community. They would respond very rapidly — within minutes or hours — to changes in MeHg concentrations in water. Phytoplankton obtain MeHg directly from water (Mason et al. 1996), and algal density can influence concentrations of MeHg in phytoplankton via biomass dilution (Pickhardt et al. 2002).

Phytoplankton are easily sampled with fine-mesh nets. However, samples require considerable processing before analysis to remove zooplankton, detritus, or other particulates. Sampling would not measurably affect target populations, even in the smallest lakes.

Little is known about MeHg in phytoplankton, particularly freshwater phytoplankton (Becker and Bigham 1995). Concentrations of MeHg in freshwater phytoplankton are related to those in water but the partitioning of MeHg between water and phytoplankton is strongly affected by concentrations of dissolved organic matter (Watras et al. 1998).

In summary, phytoplankton are present in all lakes, are trophically well defined, and are a candidate indicator for trend monitoring. They would rapidly respond to changes in the supply of MeHg. However, discrete samples of phytoplankton are not easily obtained and phytoplankton are not a recommended indicator for trend monitoring, largely because of the complexities associated with interpretation of trend data for phytoplankton. Large intra- and inter-annual variations in the MeHg content of phytoplankton would complicate the interpretation of trend data and the identification of long-term trends.

4.3.2.5 Periphyton

Periphyton are microscopic and macroscopic algae that attach to and grow on solid surfaces, such as lake bottoms, rooted aquatic vegetation, and submerged woody debris. Periphyton form part of the base of littoral food webs in lakes. Periphyton communities are taxonomically diverse and the attached communities contain other organisms, such as bacteria and zooplankton, as well as detrital material. Periphyton vary seasonally and annually in both abundance and species composition.

Periphyton are widely distributed and common; however, the trophic position of the overall community is complex and variable. Periphyton have been studied little with regard to mercury cycling but would be expected to respond rapidly to changes in the supply of MeHg. Periphyton can be directly eaten by fish, and periphyton mats may be an important site for mercury methylation in some ecosystems (Cleckner et al. 1999). Methylmercury typically comprises a very small, but quite variable, fraction of the total mercury in periphyton (Cleckner et al. 1998; Bowles et al. 2001). Periphyton are relevant neither to the public nor the policy community, because they are not consumed by people or wildlife, and their importance in the cycling of mercury is unclear. Periphyton would be relatively easy to sample but samples would contain a complex mixture of plants, small invertebrates, and detritus, complicating interpretation of the MeHg concentrations therein. Sampling would not significantly affect target populations.

In summary, periphyton are present in all lakes, easy to sample, and would respond rapidly to changes in the abundance of MeHg. Periphyton are sometimes eaten directly by fish. The diverse, complex, and variable nature of the periphyton community, however, would complicate interpretation of mercury concentrations in periphyton in a monitoring program.

4.3.3 Recommended Aquatic Biological Indicators

The criteria listed in Section 4.3.1 were applied to the selection of biological indicators to ensure their relevance and utility for assessing trends in the bioaccumulation of MeHg associated with altered loadings of mercury to aquatic systems. This evaluation, based largely on the discussion in Section 4.3.2, is summarized in Table 4.1.

We consider prey fish, piscivorous fish, and estuarine benthic invertebrates (listed in order of preference) suitable aquatic biological indicators for trend monitoring. One-year-old prey fish and piscivorous fish (analyzed individually) are the preferred biological indicators for freshwater systems. Determination of total mercury in axial muscle (preferably without skin) from piscivorous fish would indicate gradual (multi-year) trends in MeHg that are directly relevant to humans who eat sport fish. For 1-year-old prey fish, the annual sampling and analysis of either whole fish or axial muscle for total mercury would indicate annual changes in MeHg exposure in freshwater and marine ecosystems. In coastal estuaries, 1-year-old prey fish (analyzed for total mercury), as well as shellfish and macrocrustaceans (analyzed for MeHg), are recommended as biological indicators for trend monitoring. Adult piscivorous fish from heavily exploited populations or heavily fished water bodies are not recommended as biological indicators because of the potentially large confounding effects of intensive fish harvest on MeHg concentrations in such populations.

The sampling of aquatic biological indicators of different size, age, and trophic position can enhance the interpretation and understanding of temporal patterns in MeHg concentration. Concentrations of MeHg in 1-year-old prey fish will be sensitive to interannual variations in controlling factors and processes (such as mercury loadings, temperature, and hydrology), whereas older piscivorous fish will integrate and reflect temporal changes across multi-year time scales. Based on observed chronologies of mercury concentrations in aquatic biota in reservoirs during the first 30 years after flooding (Bodaly et al. 1997), we anticipate that decreases in MeHg concentrations in response to decreased atmospheric loadings would first be evident in 1-year-old prey fish, whereas decreases in concentrations in long-lived piscivores would lag a few years later.

The importance of MeHg uptake and transfer at the base of the food web is recognized and should be the subject of focused, intensive research. The utility of periphyton, phytoplankton, zooplankton, and freshwater benthic invertebrates for trend monitoring, however, is hampered by procedural and interpretational complexities associated with the large spatio-temporal variation in species assemblages, sample heterogeneity, and MeHg concentration. The utility of these groups for trend monitoring is also hampered by our very limited understanding of the processes, variables, and confounding factors that affect their bioaccumulation of MeHg.

TABLE 4.1
Recommended criteria for selection of aquatic biological indicators for monitoring and assessment of methylmercury (MeHg), and their application to candidate biological indicators

Criterion	Importance of criterion	Extent to which the aquatic biological indicator satisfies the criterion:					
		Periphyton	Phytoplankton	Zooplankton	Benthic invertebrates	Prey fish	Piscivorous fish
Relevance to the MeHg problem	To ensure relevance to human health, ecological risk, and development of policy	Low	Low	Low	Low in fresh waters; High in coastal estuaries	Medium to High	High and direct
Availability of historic MeHg data	To document the utility of the indicator and its probable reliability for detecting change in mercury loading	Low	Low	Low	Low to Medium	Low to Medium	High
Recognition of extrinsic confounding factors	To facilitate the defensible interpretation of monitoring results on MeHg	Low	Low	Medium	Medium to High	Medium to High	High
Knowledge of intrinsic co-variables	To ensure knowledge of organismal attributes that can affect MeHg concentration and complicate interpretation of results	Low	Low	Medium	Low to Medium	Medium	High

TABLE 4.1 (continued)
Recommended criteria for selection of aquatic biological indicators for monitoring and assessment of methylmercury (MeHg), and their application to candidate biological indicators

		Extent to which the aquatic biological indicator satisfies the criterion:					
Criterion	Importance of criterion	Periphyton	Phytoplankton	Zooplankton	Benthic invertebrates	Prey fish	Piscivorous fish
Broad geographic distribution	To select biotic indicators that have broad spatial coverage in a regional, national, or multi-national monitoring program	Widespread, but spatially and temporally variable assemblages	Widespread, but spatially and temporally variable assemblages	Widespread, but spatially and temporally variable assemblages	Several widespread species and genera	Several widespread species and genera	Several widespread species and genera
Important in trophic transfer of MeHg	To select biotic indicators with a significant role in the trophic transfer of MeHg in aquatic food webs	Unclear	Important	Highly important	Highly important	Highly important	Highly important
Constrained food habits or trophic position	Spatio-temporal variation in trophic position can confound and complicate interpretation of trends in MeHg concentration in the indicator	Highly constrained	Variable and complex	Variable and complex; constrained in some filter feeders	Variable and complex	Constrained	Ontogenetic shifts with increasing size; less variable in adults

		Multi-annual	Annual	Days to seasonal in small benthos; annual to multi-annual in large benthos	Hours to days	Hours to days	Hours to days
Temporal response to changes in abundance of MeHg	A slow rate of response to altered MeHg loading could increase the potential for interference by confounding factors, whereas a rapid response, coupled with high intraannual variability would hinder identification of multi-year trends						
Effects of sampling on target population or community	To reduce effects of sampling on monitored populations and biotic communities, and to limit potential confounding biotic responses to sampling	Generally nonintrusive, except in some small lakes and streams	Nonintrusive	Nonintrusive	Nonintrusive	Nonintrusive	Nonintrusive

4.4 MONITORING AND TREND ANALYSIS

We recommend that biological indicators be sampled annually to assess responses of MeHg concentrations to changes in mercury loadings and that such sampling be limited to nonmigratory species. Although sampling at less frequent intervals (e.g., every 2 or 3 years) would decrease effort and cost per monitored water body, it would substantially reduce statistical power to detect temporal trends and would delay the detection of changes in mercury concentration (Hebert and Weseloh 2003; Bignert et al. 2004).

Freshwater piscivorous fish and prey fish should be sampled in several water bodies because data from multiple sites (i.e., clusters of sites) are needed to describe trends within a given geographic area. The primary response to decreased mercury deposition within a given area may be decreased mercury concentrations in fish, yet concentrations may remain stable or increase in some water bodies because of other factors that influence mercury cycling, MeHg production, and bioaccumulation. In Minnesota, for example, analysis of state-wide trend data on total mercury in filets of standard-sized game fish from 176 lakes showed that 49% of the lakes had statistically significant ($p < 0.05$) decreases in fish-mercury concentrations, 25% had significant increases, and 26% did not change (Bruce A. Monson, Minnesota Pollution Control Agency, St. Paul, Minnesota, personal communication). Significantly more of the Minnesota lakes exhibited decreasing mercury concentrations in fish than could be attributed to chance alone (χ^2 test, $p < 0.05$). Concentrations have also declined significantly in walleyes inhabiting boreal lakes of central Canada in the last 20 to 30 years (Johnston et al. 2003), and variation among lakes in fish-mercury trends is also evident in the Canadian data (Figure 4.3).

Concentrations of MeHg in individual fish typically increase with increasing size and age, and data on the length, weight, and age of individual fish are needed to interpret trends in mercury concentration. Piscivorous fish should, therefore, be analyzed individually. Temporal trends in mercury concentrations in piscivorous fish can be biased by variation in the size or age composition of samples taken at different times. It is, therefore, useful to estimate mercury concentrations in fish of a given age or size. In Florida, mercury concentrations in axial muscle of 3-year-old large-mouth bass (*Micropterus salmoides*) have been successfully used to examine variation in fish contamination among lakes (Lange et al. 1993) and to monitor temporal trends in the Everglades (Atkeson et al. 2003). For piscivorous fish, stratified sampling by predefined length interval could be a useful framework for sampling a target population.

One widely used approach has been to sample individual fish from a target population across a range of lengths and to apply linear regression between mercury concentration and length to estimate the concentration in a fish of some standardized length; examples include 50-cm walleye and 60-cm northern pike (Parks and Hamilton 1987; Johnston et al. 2003). This procedure yields a concentration adjusted to a specific length, along with an associated confidence interval around the estimated value. Statistical power analysis of initial data can be used to estimate the sample size (number of fish) needed to detect a change of given magnitude (Exponent 2003).

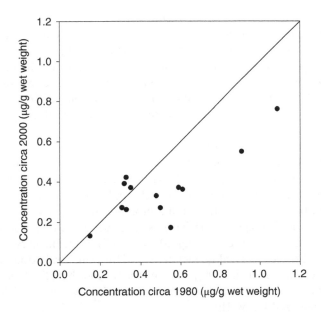

FIGURE 4.3 Recent changes in concentrations of total mercury in axial muscle of walleyes from 13 boreal lakes in northwestern Ontario and Manitoba, Canada. Standardized concentrations for 50-cm walleye sampled during 1996–2000 are plotted against standardized concentrations for fish sampled during 1977–1983 (data are for reference lakes reported in Johnston et al. 2003). Each point below the diagonal line, which has a slope of 1.0, represents a lake where the standardized concentration declined between the 2 sampling intervals.

Polynomial regression with indicator variables is another recommended statistical method for analysis of fish-mercury data. This procedure, described by Tremblay et al. (1998), allows rigorous statistical comparison of mercury-to-length relations among years and is considered superior to simple linear regression and analysis of covariance for analysis of data on mercury-length relations in fish.

For piscivorous fish, we recommend that axial muscle without skin be sampled and analyzed to avoid the additional variation that would result from differing proportions of skin and axial muscle tissue among samples. Concentrations of mercury are less in skin than in axial muscle, and the inclusion of skin in a filet can reduce the concentration of total mercury or MeHg in the sample by 5 to 12 percent (Glass et al. 1998; Serdar et al. 2001).

The historic data on total mercury in piscivorous fish are generally considered valid, and it could be useful to perform trend monitoring in carefully selected fresh waters for which historic data are acceptable and sufficient for statistical comparisons. In the late 1960s and early 1970s, the discovery of mercury-contaminated fish prompted widespread surveillance and monitoring of mercury in fish. The earliest surveys focused largely on surface waters that received mercury in polluted wastewater, but often included fish sampled from presumed "reference sites" that did not receive mercury from industrial point-source discharges. The surveillance of mercury in fish was expanded spatially with growing awareness of the importance of atmospheric deposition as a potentially significant source of mercury in surface waters lacking on-site anthropogenic sources (Wiener et al. 2003). Historic mercury data for fish from surface waters that did not receive point-source discharges of mercury may be useful for assessing decadal trends related to atmospheric deposition. Historic fish-mercury data from analyses of composite samples (containing subsamples from multiple fish), however, has limited statistical utility for trend analysis (Exponent 2003).

We recommend that prey fish be sampled in early spring or fall, when growth is slow and temporal variation in mercury concentration is less pronounced, to facilitate comparisons among years and surface waters. Both axial muscle and whole fish have been successfully used in research applications, and analysis of either matrix would be suitable in a monitoring program for mercury. Determination of total mercury would be appropriate for muscle tissue, given that nearly all ($\geq$95%) of the mercury in fish muscle is MeHg. Most (>70%) of the mercury in whole prey fish is MeHg (Francesconi and Lenanton 1992; Hill et al. 1996; Bodaly and Fudge 1999; Hammerschmidt et al. 1999), and we recommend that all whole prey fish be analyzed for total mercury and that a subset (from 5 to 20% of the fish, selected at random) be analyzed for MeHg.

If whole prey fish are analyzed, data should be reported as whole-body concentration and burden. Mercury *burden*, defined as the total mass of mercury accumulated in a whole fish, is calculated as the product of body weight and whole-body concentration. In prey fish of known age (e.g., age-1), mercury burden is an ecologically relevant measure of bioaccumulation, representing the mass of mercury accumulated by a fish during its residence in a monitored water body. The burden also represents the mass of mercury that a predator would ingest when eating the prey fish. Prey fish should be analyzed individually, and ancillary measurements include the total length (to nearest millimeter), fresh weight (to 0.01 g), and age of individual fish. Age can be estimated by examination of scales or other bony structures (DeVries and Frie 1996).

The trophic position of candidate benthic invertebrates for trend monitoring should be known or estimated so that samples with equivalent trophic status can be analyzed and compared. Stable-isotope methods are available for estimating trophic position (Fry 1991; Broman et al. 1992) and for assessing the influence of trophic position on mercury concentrations (Cabana and Rasmussen 1994; Kidd et al. 1995). The application of $\delta^{15}N$ and associated metrics in monitoring programs would enhance our ability to understand trends and strengthen inferences concerning linkages between trends in biotic concentrations and mercury loadings. The corrections and enhancements of Phillips (2001) and Phillips and Gregg (2001) should be considered when using data on stable isotopes to estimate the trophic status of benthic species.

The technique that has been widely applied for analyzing total mercury in aquatic biota since the late 1960s (cold vapor atomic absorption spectrophotometry) remains a valid analytical method. Accordingly, we infer that most of the historical data for total mercury in fish tissues are valid. Moreover, the historical data on total mercury concentrations in fish tissues provide defensible estimates of prior MeHg concentrations.

This is in contrast to most data collected before 1990 on total mercury and MeHg in water; most of these early data for water are not considered valid because sensitive analytical methods were not sufficiently developed and samples were typically contaminated during handling. There are some historical data on total mercury in aquatic invertebrates, but comparatively few estimates of MeHg in historic samples. Thus, the historical mercury record for piscivorous fish in North America is now about 35 years long, whereas the much sparser historical record for MeHg in water and aquatic biota in lower trophic levels is much shorter, perhaps 10 to 20 years in most regions. In comparison, the historic record of total-mercury deposition that can be examined by analyses of dated cores of sediment, peat, and glacial ice spans 2 to several centuries (Engstrom and Swain 1997; Martínez-Cortizas et al. 1999; Bindler et al. 2001; Schuster et al. 2002; Shotyk et al. 2003).

4.5 ANCILLARY DATA

The defensible interpretation of trend data in a biological monitoring program for mercury will be enhanced by the concurrent collection of relevant information on monitored airsheds, watersheds, and surface waters (see Chapters 2 and 3). The interpretation of fish-mercury data and the identification of factors causing temporal shifts in MeHg concentrations in aquatic biota would be strengthened by co-locating trend monitoring for airsheds, watersheds, and biological indicators at intensive study sites (Driscoll et al. this volume (Chapter 2), Krabbenhoft et al. this volume (Chapter 3)). Important metrics for monitored airsheds and watersheds include annual atmospheric deposition of total mercury, MeHg, and sulfate; interannual variations in rainfall and temperature; watershed area; land cover, abundance of hydrologically connected wetlands; and annual export of total mercury and MeHg from catchments to monitored surface waters. For lakes, key limnological data include lake morphometry (area, maximum depth, mean depth, percent littoral area); physicochemical characteristics of water (pH, dissolved organic carbon, sulfate, total suspended solids, chlorophyll, acid neutralizing capacity, color, and phosphorus); depth profiles of temperature and dissolved oxygen during summer stratification; hydrologic type (e.g., seepage or drainage lake); and concentrations of total mercury and MeHg in oxic water (preferably sampled when lakes are mixed). Information on the depositional chronology of total mercury in dated sediment cores would be useful for lakes with a multi-decadal, historic record of mercury concentrations in fish. Ancillary site-specific biological data include harvest of fish and shellfish; results of biotic and fishery surveys; and the age, growth rate, condition, sex, and trophic position (inferred from stable-isotope and dietary analyses) of aquatic biological indicators.

A number of recent experiments have shown that the reproduction and fitness of fish can be adversely affected by exposure to environmentally realistic concentrations of MeHg (Fjeld et al. 1998; Latif et al. 2001; Hammerschmidt et al. 2002; Drevnick and Sandheinrich 2003). In cases where trend monitoring reveals high MeHg concentrations in biological indicator organisms, subsequent field and laboratory studies should be considered to assess toxicological effects.

4.6 INTERPRETATION OF TREND-MONITORING DATA

Several biogeochemical processes, environmental factors, and human disturbances can influence the production of MeHg, its abundance in surface waters, and its concentrations in aquatic biota (Jackson 1998; Lucotte et al. 1999c; Benoit et al. 2003; Wiener et al. 2003). Thus, MeHg concentrations often vary spatially in aquatic organisms of the same species and size, even among nearby water bodies that receive similar atmospheric loadings of mercury (Wiener et al. 2003). The bioaccumulative responses of aquatic organisms to altered external loadings of mercury can, therefore, be expected to vary substantially among surface waters. Even within a given water body, internal ecosystem dynamics, anthropogenic activities unrelated to mercury loadings, and long-term external factors could confound or obscure responses in MeHg bioaccumulation to altered mercury loadings (e.g., Sorensen et al. 2005). Similarity of pertinent biogeochemical, environmental, and human factors affecting MeHg concentrations in aquatic organisms — among sampling intervals in a trend-monitoring program — would enhance our ability to detect responses to altered mercury loadings. Such similarity should not, however, be expected or assumed in a trend-monitoring program.

4.6.1 Sources of Variation and Potential Confounding Factors

Here we examine sources of variation and potential confounding factors that could obscure temporal changes in MeHg concentrations in aquatic biota in response to altered loadings of mercury. General sources of variation and potential confounding factors include interannual variations and dynamics of the monitored ecosystems, direct influences of anthropogenic and natural activities, and long-term external factors. Knowledge of such factors is prerequisite to discerning trends in MeHg concentrations in aquatic biota associated with altered loadings of mercury to aquatic systems. Awareness of potential confounding factors is also essential for minimizing or statistically accounting for their effects in a trend-monitoring program (see Chapter 3).

1) *Methylation and demethylation.* The methylation of inorganic mercury and demethylation of MeHg are key processes influencing the abundance of MeHg in food webs supporting the production of shellfish and fish (Benoit et al. 2003; Wiener et al. 2003). Rates of methylation and demethylation are affected by several variables, including temperature, sulfur cycling, light penetration, and the amount and quality of available organic matter (Miskimmin et al. 1992; Allan et al. 2001; Benoit et al. 2003; Kainz et al. 2003; Rencz et al. 2003). Such controlling variables can vary in response to natural events or conditions, such as water-level fluctuations and flooding, atypically cold or warm weather, increased turbidity caused by resuspension of bottom sediment, or varying abundances of allochthonous and autochthonous organic matter.

2) *Food-web and trophic dynamics.* Organisms at the base of aquatic food webs play a key role in the transfer of MeHg to upper trophic levels (Jackson and Harvey 1993; Plourde et al. 1997; Tremblay and Lucotte 1997; Bodaly and Fudge 1999). Within a given water body, interannual

variations in climatic and environmental conditions can affect the composition and abundance of pelagic and benthic invertebrates. Such variation in invertebrate assemblages can alter MeHg bioaccumulation in small and large fish at the end of the growing season (Parkman and Meili 1993; Montgomery et al. 2000; Gorski et al. 2003; Kainz et al. 2003).

Most fish are opportunistic organisms whose diets and trophic position can vary with food availability and size. Variation in environmental conditions among years (including water levels during spawning, availability of spawning sites, winter harshness, and summer oxygen depletion) can alter trophic structure of the fish community and the metabolism of fish, thereby affecting concentrations of MeHg in fish tissue (Lockhart et al. 1972; Meili 1991; Harris and Bodaly 1998; Scheuhammer and Graham 1999; Stafford and Haines 2001; Gorski et al. 2003). Interannual variation in growth rates of fish could influence the size-standardized mercury concentrations that are used to assess temporal trends in MeHg accumulation (Tremblay et al. 1998; Johnston et al. 2003; Sonesten 2003).

3) *Landscape disturbance.* Forest soils contain large inventories of mercury (Roulet et al. 1998; Lucotte et al. 1999a; Rencz et al. 2003) that can be mobilized into aquatic systems by disturbance of the forest landscape. Logging and fire can expose soil and increase the transport of mercury from soil to aquatic systems, a situation observed in northern boreal forest (Garcia and Carignan 1999; Porvari et al. 2003) and in Amazonian rainforest (Roulet et al. 1999, 2000). Agricultural practices, urban development, and road construction can also contribute to soil erosion and increased transport of mercury in soil from terrestrial to aquatic environments. Human disturbances of a watershed could also indirectly affect the production and bioaccumulation of MeHg in an aquatic system, by altering hydrologic pathways, water levels, nutrient inputs, primary production, light penetration, and trophic status.

4) *Fishing intensity.* Temporal variation in fishing intensity and harvest in a monitored water body could confound the analysis of trends in mercury concentrations in adult piscivorous fish. In Finnish and Scandinavian lakes, mercury concentrations in fish of given length declined notably a few years after intensive fishing had removed substantial numbers of the large adult fish (Göthberg 1983; Verta 1990). In experimental boreal lakes in Canada, the inventories of mercury in the water column and surface sediment were reduced little by intensive fishing (Surette et al. 2006). Intensive fishing may have affected the structure of the zooplankton communities (Masson and Tremblay 2002); however, the diets of the studied fishes did not change significantly after intensive fishing (Doire et al. 2002). Intensive fishing, however, significantly increased the growth rate of large predatory fish (Simoneau et al. 2005) and markedly decreased MeHg concentrations in fish (Surette et al. 2003).

5) *Exotic species.* The voluntary or accidental introduction of exotic species may influence the trophic structure of aquatic systems (Hrabik et al. 1998), thereby affecting MeHg concentrations in aquatic organisms in upper

trophic levels. Johnston et al. (2003), however, found that the introduction of rainbow smelt (*Osmerus mordax*) did not significantly affect concentrations of mercury in native predatory fish in lakes of central Canada.

6) *Reservoir impoundment.* The bioaccumulation of MeHg in aquatic organisms is greatly increased by the flooding of vegetated wetland or upland terrestrial habitats. This has been well documented in newly flooded reservoirs in temperate, boreal, and tropical regions (Bodaly et al. 1984; Johnston et al. 1991; Porvari 1995; Plourde et al. 1997; Seda and Kubecka 1997; Tremblay and Lucotte 1997; Lucotte et al. 1999b; Verdon and Tremblay 1999; Montgomery et al. 2000). Flooding of new reservoirs greatly increases the rate of mercury methylation, causing rapid increases (days to a few weeks) in concentrations of MeHg in water, zooplankton, and prey fish (Paterson et al. 1998; Bodaly and Fudge 1999). Concentrations of MeHg in the axial muscle of adult piscivorous fish can increase as much as 10-fold relative to pre-flood or reference values and remain elevated for decades after initial impoundment (Porvari 1998; Schetagne and Verdon 1999). New reservoirs also export MeHg, greatly increasing the concentrations of mercury in aquatic biota and fish inhabiting downstream aquatic environments (Johnston et al. 1991; Schetagne and Verdon 1999). In older reservoirs, concentrations of mercury in prey fish can fluctuate substantively in response to water-level fluctuations (Sorensen et al. 2005).

7) *Climate change.* Climate change can influence an array of environmental variables that directly and indirectly affect the biogeochemical cycling of mercury. The microbial methylation of mercury is temperature sensitive, with increasing production and concentrations of MeHg in temperate aquatic ecosystems at higher temperature (Bodaly et al. 1993). Climate change can modify the hydrological cycle, thereby altering mercury transport, transformations, and trophic structure in aquatic systems. Changes in the intensity of ultraviolet (UV) radiation at the water surface could influence both the production and photodemethylation (destruction) of MeHg in surface waters (Sellers et al. 1996, 2001; Lean and Siciliano 2003).

8) *Acidity and sulfate content of wet deposition.* Temporal variation in the acidity and sulfate content of wet deposition could modify the accumulation of MeHg in aquatic biota, given that numerous interactions between mercury transformations, MeHg uptake, sulfate, and aqueous pH have been reported (Spry and Wiener 1991; Odin et al. 1994; Frost et al. 1999; Scheuhammer and Graham 1999; Kelly et al. 2003). The biogeochemistry of mercury in waters with low acid neutralizing capacity and in methylating wetland environments could be affected by long-term chemical changes in wet deposition.

4.6.2 STEPS TO CONSTRAIN CONFOUNDING FACTORS AND ENHANCE INTERPRETATION

Monitoring programs should be designed to limit the effects of confounding factors that impair our ability to discern both trends in MeHg concentrations in aquatic biota and their relation to altered atmospheric loadings of mercury. Adherence to the guidelines below should eliminate some of the pitfalls that can complicate interpretation of trend data on MeHg in aquatic biota. The application of these guidelines during the development of a monitoring program requires compilation of appropriate information on candidate study sites (surface waters and their watersheds) and their resident aquatic biota.

1) *Exclude very contaminated surface waters during site selection.* Concentrations of MeHg in biota inhabiting surface waters historically contaminated by wastes from industrial point-source discharges or historic mining activities can remain substantially elevated for decades (Latif et al. 2001; Wiener et al. 2003). In new reservoirs, the concentrations of MeHg in piscivorous fish remain elevated for several years after initial flooding, declining gradually over 2 to 3 decades (Bodaly et al. 1997). A trend-monitoring program intended to assess responses to changes in atmospheric loading should not include surface waters that are recovering from prior on-site elevation in abundance of either total mercury or MeHg.

2) *Exclude surface waters on disturbed watersheds.* Surface waters in watersheds that have recently undergone disturbance (e.g., fire or notable human development) or in watersheds where disturbance is anticipated during monitoring (e.g., logging or changing land use) should be avoided during site selection. This requires access to information on prior land use, as well as land-management plans. Given this, semi-remote surface waters in protected areas on state, provincial, or federal lands, such as national parks, would be appropriate sites for trend monitoring.

3) *Exclude surface waters subjected to intensive or highly variable fishing pressure.* Surface waters that are subjected to intensive fishing pressure or that are expected to undergo highly variable commercial or recreational harvest of fish or shellfish during monitoring should be excluded during site selection.

4) *Employ nonlethal methods when sampling small populations of piscivores.* In very small lakes or streams that support small populations, the application of procedures for nonlethal sampling (e.g., Baker et al. 2004; Peterson et al. 2005) may be advisable to avoid potential effects of removal sampling on MeHg concentrations in target populations of piscivorous fish. This determination requires advance knowledge of bioindicator organisms present at a candidate monitoring site, the approximate number (sample size) of an indicator organism to be taken during each sampling event, and the approximate size of the target population being sampled.

5) *Include multiple water bodies (clusters of sites) within each monitored geographic area.* Within a geographic area, the chosen biological indicators should be sampled in several water bodies of a given type (i.e., clusters of lakes or streams), because data from several sites will probably be needed to identify the *overall* trend within the area (Section 4.4). The direction of temporal trends can differ among individual water bodies because of spatio-temporal variation in other factors that influence mercury

cycling, net methylation, bioaccumulation, and concentrations in organisms. Statistical analyses should focus first on detecting temporal trends within individual sites and second on evaluating the overall direction of change across the cluster of sites within a monitored geographic area. Descriptive and exploratory statistical analyses should characterize monitored ecosystems that exhibit different trends (e.g., significant decrease, no change, or significant increase in MeHg concentrations) within a given area, as an initial step toward identifying potential causes of the observed differences in trends.

6) *Monitor bioaccumulation in some lakes with minimal watershed influence.* Delineating between watershed-derived mercury and atmospherically derived mercury is problematic in a biological trend-monitoring program. Wetlands can be particularly important sites of MeHg production and significant sources of total mercury, MeHg, and organic matter for adjoining surface waters (Hurley et al. 1995; St. Louis et al. 1996; Curtis 1998; Sellers et al. 2001). Concentrations of MeHg in aquatic biota should respond most rapidly to altered mercury deposition in atmospherically dominated systems, such as perched seepage lakes, which lack streams and receive little or no inflow of surface or ground water from the surrounding terrestrial environment. The production of MeHg and its bioaccumulation in such seepage lakes are therefore influenced little by watershed characteristics and processes (Watras et al. 1994; Hrabik and Watras 2002; Watras et al. 2002; Wiener et al. 2003).

7) *Ensure that contemporary and historic monitoring data are valid and comparable.* Mercury data from different biological monitoring programs often are not directly comparable because of variation in methods for sampling, sample preparation, and analyses. Contemporary monitoring efforts that intend to use historic data on biological indicators should take steps to ensure that 1) the historic data are valid and statistically useful and 2) the methods applied in contemporary and historic monitoring produce comparable data. Quality assurance procedures should be implemented to document the accuracy and ensure the reliability of analytical measurements. Increased uniformity in methods for sampling, sample preparation, chemical analyses, database management, and statistical summarization of data would facilitate comparison and synthesis of mercury data across geopolitical boundaries.

8) *Continue monitoring for a sufficient time frame.* Temporal records of mercury concentrations in aquatic biota should be sufficiently long to detect decade-scale trends that could be obscured by short-term fluctuations caused by interannual variability. This requires a commitment to sustained allocation of resources at the onset of a trend-monitoring program, for a recommended time frame of 20 to 30 years. Existing monitoring or research programs may include suitable candidate sites for monitoring, for which reliable historic data on mercury and pertinent ancillary variables are available.

Additional detailed guidance concerning the development of monitoring programs for bioaccumulative contaminants is available from the U.S. Environmental Protection Agency (2000) and from other chapters in this book (Chapters 2, 3, and 5). Provision of statistical guidance relevant to sampling design and data analyses is beyond the scope of this chapter, but such information is available from other sources (Gilbert 1987; Jorgensen and Pedersen 1994; Tremblay et al. 1998; Conquest 2000; U.S. Environmental Protection Agency 2000).

ACKNOWLEDGMENTS

The lead author (JGW) was supported by a Wisconsin Distinguished Professorship jointly funded by the University of Wisconsin System, the University of Wisconsin–La Crosse (UW–L) Foundation, and the UW–L College of Science and Health during the preparation of this chapter. We thank Michael Paterson (Fisheries and Oceans Canada, Winnipeg, Manitoba) for allowing the use of his unpublished data on methylmercury in zooplankton; Bruce Monson (Minnesota Pollution Control Agency, St. Paul) for sharing results from his trend analysis of mercury in game fish from Minnesota lakes; and Thomas Johnston (Fisheries and Oceans Canada, Burlington, Ontario) for providing trend data on mercury in walleyes from lakes in Ontario and Manitoba. Constructive reviews of a draft chapter were provided by Robin Stewart, Mark Brigham, Michael Murray, Reed Harris, and Tamara Saltman.

REFERENCES

Allan CJ, Heyes A, Roulet NT, St Louis VL, Rudd JWM. 2001. Spatial and temporal dynamics of mercury in Precambrian Shield upland runoff. Biogeochemistry 52:13–40.

Andersson P, Borg H, Kaerrhage P. 1995. Mercury in fish muscle in acidified and limed lakes. Water Air Soil Pollut 80:889–892.

Atkeson T, Axelrad D, Pollman C, Keeler G. 2003. Integrating atmospheric mercury deposition and aquatic cycling in the Florida Everglades: integrated summary. Tallahassee (FL): Florida Department of Environmental Protection, p. 75–90.

Back RC, Visman V, Watras CJ. 1995. Microhomogenization of individual zooplankton species improves mercury and methylmercury determinations. Can J Fish Aquat Sci 52:2470–2475.

Back RC, Watras CJ. 1995. Mercury in zooplankton of northern Wisconsin lakes: taxonomic and site-specific trends. Water Air Soil Pollut 80:931–938.

Baker RF, Blanchfield PJ, Paterson MJ, Flett RJ, Wesson L. 2004. Evaluation of non-lethal methods for the analysis of mercury in fish tissue. Trans Am Fish Soc 133:568–576.

Barr JF. 1996. Aspects of common loon (*Gavia immer*) feeding biology on its breeding ground. Hydrobiologia 321:119–144.

Becker DS, Bigham GN. 1995. Distribution of mercury in the aquatic food web of Onondaga Lake, New York. Water Air Soil Pollut 80:563–571.

Becker GC. 1983. Fishes of Wisconsin. Madison (WI): University of Wisconsin Press. 1052 p.

Benoit JM, Fitzgerald WF, Damman AWH. 1998. The biogeochemistry of an ombrotrophic bog: evaluation of use as an archive of atmospheric mercury deposition. Environ Res (Sect A) 78:118–133.

Benoit J, Gilmour C, Heyes A, Mason RP, Miller C. 2003. Geochemical and biological controls over methylmercury production and degradation in aquatic ecosystems. In: Chai Y, Braids OC, editors, Biogeochemistry of environmentally important trace elements. Washington, DC: American Chemical Society, p. 262–297.

Bignert A, Riget F, Braune B, Outridge P, Wilson S. 2004. Recent temporal trend monitoring of mercury in Arctic biota — how powerful are the existing data sets? J Environ Monitoring 6:351–355.

Bindler R, Renberg I, Appleby PG, Anderson NJ, Rose NL. 2001. Mercury accumulation rates and spatial patterns in lake sediments from West Greenland: a coast to ice margin transect. Environ Sci Technol 35:1736–1741.

Bloom NS. 1992. On the chemical form of mercury in edible fish and marine invertebrate tissue. Can J Fish Aquat Sci 49:1010–1017.

Bodaly RA, Fudge RJP. 1999. Uptake of mercury by fish in an experimental boreal reservoir. Arch Environ Contam Toxicol 37:103–109.

Bodaly RA, Hecky RE, Fudge RJP. 1984. Increases in fish mercury levels in lakes flooded by the Churchill River diversion, northern Manitoba. Can J Fish Aquat Sci 41:682–691.

Bodaly RA, Rudd JWM, Fudge RJP, Kelly CA. 1993. Mercury concentrations in fish related to size of remote Canadian shield lakes. Can J Fish Aquat Sci 50:980–987.

Bodaly RA, St. Louis VL, Paterson MJ, Fudge RJP, Hall BD, Rosenberg DM, Rudd JWM. 1997. Bioaccumulation of mercury in the aquatic food chain in newly flooded areas. In: Sigel A, Sigel H, editors, Metal ions in biological systems, Vol. 34: Mercury and its effects on environment and biology. New York (NY): Marcel Dekker Inc., p. 259–287.

Bowles KC, Apte SC, Maher WA, Kawei M, Smith R. 2001. Bioaccumulation and biomagnification of mercury in Lake Murray, Papua New Guinea. Can J Fish Aquat Sci 58:888–897.

Broman D, Naf C, Rolff C, Zebuhr Y, Fry B, Hobbie J. 1992. Using ratios of stable nitrogen isotopes to estimate bioaccumulation and flux of polychlorinated dibenzo-p-dioxins (PCDDs) and dibenzofurans (PCDFs) in two food chains from the northern Baltic. Environ Toxicol Chem 11:331–345.

Brown SS, Gaston GR, Rakocinski CF, Heard RW. 2000. Effects of sediment contaminants and environmental gradients on macrobenthic community trophic structure in Gulf of Mexico estuaries. Estuaries 23:411–424.

Cabana G, Rasmussen JB. 1994. Modelling food chain structure and contaminant bioaccumulation using stable nitrogen isotopes. Nature 372:255–257.

Cizdziel JV, Hinners TA, Pollard JE, Heithmar EM, Cross CL. 2002. Mercury concentrations in fish from Lake Mead, USA, related to fish size, condition, trophic level, location, and consumption risk. Arch Environ Contam Toxicol 43:309–317.

Clarkson TW. 2002. The three modern faces of mercury. Environ Health Perspect 110(Suppl 1):11–23.

Cleckner LB, Garrison PG, Hurley JP, Olson ML, Krabbenhoft DP. 1998. Trophic transfer of methyl mercury in the northern Florida Everglades. Biogeochemistry 40:347–361.

Cleckner LB, Gilmour CC, Krabbenhoft DP, Hurley JP. 1999. Mercury methylation in periphyton of the Florida Everglades. Limnol Oceanogr 44:1815–1825.

Colby PJ, McNicol RE, Ryder RA. 1979. Synopsis of biological data on the walleye *Stizostedion vitrem vitreum* (Mitchill 1818). Rome, Italy: Food and Agriculture Organization of the United Nations. FAO Fisheries Synopsis No 119. 139 p.

Coleman FC, Figueira WF, Ueland JS, Crowder LB. 2004. The impact of United States recreational fisheries on marine fish populations. Science 305:1958–1960.

Conquest LL. 2000. Environmental monitoring: investigating associations and trends. In: Sparks T, editor, Statistics in ecotoxicology. New York (NY): John Wiley & Sons, p. 179–210.

Cope WG, Wiener JG, Rada RG. 1990. Mercury accumulation in yellow perch in Wisconsin seepage lakes: relation to lake characteristics. Environ Toxicol Chem 9:931–940.

Curtis PJ. 1998. Climatic and hydrologic control of DOM concentration and quality in lakes. In: Hessen DO, Tranvik LJ, editors, Aquatic humic substances: ecology and bio-geochemistry. Berlin, Germany: Springer-Verlag, p. 93–105.

DeVries DR, Frie RV. 1996. Determination of age and growth. In: Murphy BR, Willis DW, editors, Fisheries techniques, 2nd ed. Bethesda (MD): American Fisheries Society, p. 483–512.

Doire J, Lucotte M, Fortin R, Verdon R. 2002. Influence of intensive fishing on fish diet in natural lakes from northern Québec: use of stable nitrogen and carbon isotopes (abstract). American Society of Limnology and Oceanography, Summer Meeting, Victoria, BC, Canada, June 10–14, 2002. http://www.aslo.org/meetings/victoria2002/archive/409.html

Downs SG, Macleod CL, Lester JN. 1998. Mercury in precipitation and its relation to bioaccumulation in fish: a literature review. Water Air Soil Pollut 108:149–187.

Drevnick PE, Sandheinrich MB. 2003. Effects of dietary methylmercury on reproductive endocrinology of fathead minnows. Environ Sci Technol 37:4390–4396.

Engstrom DR, Swain EB. 1997. Recent declines in atmospheric mercury deposition in the Upper Midwest. Environ Sci Technol 31:960–967.

Environmental Council of States and Clean Air Network. 2001. Mercury in the environment: states respond to the challenge. Washington, DC: Environmental Council of States. 70 p. http://www.sso.org/ecos/projects/Mercury/ECOS2%20State%20Hg%20Activities.pdf.

Exponent. 2003. Fish contaminant monitoring program: review and recommendations prepared for Michigan Department of Environmental Quality, Water Division, Lansing, MI, Doc. No. 8601969.001 0501 0103 BH29. http://www.deq.state.mi.us/documents/deq-wd-fcmp-fcmpfinal.pdf

Fitzgerald WF, Engstrom DR, Mason RP, Nater EA. 1998. The case for atmospheric mercury contamination in remote areas. Environ Sci Technol 32:1–7.

Fjeld E, Haugen TO, Vøllestad LA. 1998. Permanent impairment in the feeding behavior of grayling (Thymallus thymallus) exposed to methylmercury during embryogenesis. Sci Total Environ 213:247–254.

Francesconi KA, Lenanton RCJ. 1992. Mercury contamination in a semi-enclosed marine embayment: organic and inorganic mercury content of biota, and factors influencing mercury levels in fish. Mar Environ Res 33:189–212.

Frost TM, Montz PK, Kratz TK, Badillo T, Brezonik PL, Gonzalez MJ, Rada RG, Watras CJ, Webster KE, Wiener JG, Williamson CE, Morris DP. 1999. Multiple stresses from a single agent: diverse responses to the experimental acidification of Little Rock Lake, Wisconsin. Limnol Oceanogr 44(3, part 2):784–794.

Fry B. 1991. Stable isotope diagrams of freshwater food webs. Ecology 72:2293–2297.

Garcia E, Carignan R. 1999. Impact of wildfire and clear-cutting in the boreal forest on methylmercury in zooplankton. Can J Fish Aquat Sci 56:339–345.

Gilbert RO. 1987. Statistical methods for environmental pollution monitoring. New York (NY): Van Nostrand Reinhold Company. 320 p.

Gilmour CC, Riedel GS, Ederington MC, Bell JT, Benoit JM, Gill GA, Stordal MC. 1998. Methylmercury concentrations and production rates across a trophic gradient in the northern Everglades. Biogeochemistry 40:327–345.

Glass GE, Sorensen JA, Rapp GR Jr. 1998. Mercury deposition and lake quality trends. Report to the Minnesota Pollution Control Agency, St. Paul, MN.

Gorski PR, Cleckner LB, Hurley JP, Sierzen ME, Armstrong DE. 2003. Factors affecting enhanced mercury bioaccumulation in inland lakes of Isle Royale National Park, USA. Sci Total Environ 304:327–348.

Gorski PR, Lathrop RC, Hill SD, Herrin RT. 1999. Temporal mercury dynamics and diet composition in the mimic shiner. Trans Am Fish Soc 128:701–712.

Göthberg A. 1983. Intensive fishing — a way to reduce the mercury level in fish. Ambio 12:259–261.

Grieb TM, Driscoll CT, Gloss SP, Schofield CL, Bowie GL, Porcella DB. 1990. Factors affecting mercury accumulation in fish in the upper Michigan peninsula. Environ Toxicol Chem 9:919–930.

Haines TA, Komov VT, Matey VE, Jagoe CH. 1995. Perch mercury content is related to acidity and color of 26 Russian lakes. Water Air Soil Pollut 85:823–828.

Hall BD, Bodaly RA, Fudge RJP, Rudd JWM, Rosenberg DM. 1997. Food as the dominant pathway of methylmercury uptake by fish. Water Air Soil Pollut 100:13–24.

Hammerschmidt CR, Sandheinrich MB, Wiener JG, Rada RG. 2002. Effects of dietary methylmercury on reproduction of fathead minnows. Environ Sci Technol 36:877–883.

Hammerschmidt CR, Wiener JG, Frazier BE, Rada RG. 1999. Methylmercury content of eggs in yellow perch related to maternal exposure in four Wisconsin lakes. Environ Sci Technol 33:999–1003.

Harris RC, Bodaly RA. 1998. Temperature, growth and dietary effects on fish mercury dynamics in two Ontario lakes. Biogeochemistry 40:175–187.

Hebert CE, Weseloh DVC. 2003. Assessing temporal trends in contaminants from long-term avian monitoring programs: the influence of sampling frequency. Ecotoxicology 12:141–151.

Heinz GH. 1996. Mercury poisoning in wildlife. In: Fairbrother A, Locke LN, Hoff GL, editors, Noninfectious diseases of wildlife, 2nd ed. Ames (IA): Iowa State University Press, p. 118–127.

Henry EA, Dodge-Murphy LJ, Bigham GN, Klein SM, Gilmour CC. 1995. Total mercury and methylmercury mass balance in an alkaline, hypereutrophic urban lake (Onondaga Lake, NY). Water Air Soil Pollut 80:509–518.

Herrin RT, Lathrop RC, Gorski PR, Andren AW. 1998. Hypolimnetic methylmercury and its uptake by plankton during fall destratification: a key entry point of mercury into lake food chains? Limnol Oceanogr 43:1476–1486.

Hill WR, Stewart AJ, Napolitano GE. 1996. Mercury speciation and bioaccumulation in lotic primary producers and primary consumers. Can J Fish Aquat Sci 53:812–819.

Hrabik TR, Magnuson JJ, McLain AS. 1998. Predicting the effects of rainbow smelt on native fishes in small lakes: evidence from long-term research on two lakes. Can J Fish Aquat Sci 55:1364–1371.

Hrabik TR, Watras CJ. 2002. Recent declines in mercury concentration in a freshwater fishery: isolating the effects of de-acidification and decreased atmospheric mercury deposition in Little Rock Lake. Sci Total Environ 297:229–237.

Hurley JP, Benoit JM, Babiarz CL, Shafer MM, Andren AW, Sullivan JR, Hammond R, Webb DA. 1995. Influences of watershed characteristics on mercury levels in Wisconsin waters. Environ Sci Technol 29:1867–1875.

Hutchings JA, Reynolds JD. 2004. Marine fish population collapses: consequences for recovery and extinction risk. BioScience 54:297–309.

Jackson DA, Harvey HH. 1993. Fish and benthic invertebrates: community concordance and community-environment relationships. Can J Fish Aquat Sci 50:2641–2651.

Jackson TA. 1997. Long-range atmospheric transport of mercury to ecosystems, and the importance of anthropogenic emissions — a critical review and evaluation of the published evidence. Environ Rev 5:99–120.

Jackson TA. 1998. Mercury in aquatic ecosystems. In: Langston WJ, Bebianno MJ, editors, Metal metabolism in aquatic environments. London, UK: Chapman & Hall, p. 77–158.

Johansson K, Aastrup M, Andersson A, Bringmark L, Iverfeldt Å. 1991. Mercury in Swedish forest soils and waters — assessment of critical load. Water Air Soil Pollut 56:267–281.

Johnston TA, Bodaly RA, Matias JA. 1991. Predicting fish mercury levels from physical characteristics of boreal reservoirs. Can J Fish Aquat Sci 48:1468–1475.

Johnston TA, Leggett WC, Bodaly RA, Swanson HK. 2003. Temporal changes in mercury bioaccumulation by predatory fishes of boreal lakes following the invasion of an exotic forage fish. Environ Toxicol Chem 22:2057–2062.

Jorgensen LA, Pedersen B. 1994. Trace metals in fish used for time trend analysis and as environmental indicators. Marine Pollut Bull 28:235–243.

Kainz M, Lucotte M, Parrish CC. 2003. Relationships between organic matter composition and methylmercury content of offshore and carbon-rich littoral sediments in an oligotrophic lake. Can J Fish Aquat Sci 6:888–896.

Kelly CA, Rudd JWM, Holoka MH. 2003. Effect of pH on mercury uptake by an aquatic bacterium: implications for Hg cycling. Environ Sci Technol 37:2941–2946.

Kidd KA, Hesslein RH, Fudge RJP, Hallard KA. 1995. The influence of trophic level as measured by $\delta^{15}N$ on mercury concentrations in freshwater organisms. Water Air Soil Pollut 80:1011–1015.

Kraepiel AML, Keller K, Chin HB, Malcolm EG, Morel FMM. 2003. Sources and variations of mercury in tuna. Environ Sci Technol 37:5551–5558.

Lamborg CH, Fitzgerald WF, Damman AWH, Benoit JM, Balcom PH, Engstrom DR. 2002. Modern and historic atmospheric mercury fluxes in both hemispheres: global and regional mercury cycling implications. Global Biogeochem Cycles 16(4):1104.

Lange TR, Royals HE, Connor LL. 1993. Influence of water chemistry on mercury concentration in largemouth bass from Florida lakes. Trans Am Fish Soc 122:74–84.

Latif MA, Bodaly RA, Johnston TA, Fudge RJP. 2001. Effects of environmental and maternally derived methylmercury on the embryonic and larval stages of walleye (*Stizostedion vitreum*). Environ Pollut 111:139–148.

Lean DRS, Siciliano SD. 2003. Production of methylmercury by solar radiation. J Phys IV, Proc 12th Internat Conf on Heavy Metal in the Environment 107:743–747.

Lin CJ, Cheng MD, Schroeder WH. 2001. Transport patterns and potential sources of total gaseous mercury measured in Canadian high Arctic in 1995. Atmos Environ 35:1141–1154.

Lindestroem L. 2001. Mercury in sediment and fish communities of Lake Vaenern, Sweden: recovery from contamination. Ambio 30:538–544.

Lockhart WL, Uthe JF, Kenney AR, Mehrle PM. 1972. Methylmercury in northern pike (*Esox lucius*): distribution, elimination, and some biochemical characteristics of contaminated fish. J Fish Res Board Can 29:1519–1523.

Lockhart WL, Wilkinson P, Billeck BN, Danell RA, Hunt RV, Brunskill GJ, DeLaronde J, St. Louis V. 1998. Fluxes of mercury to lake sediments in central and northern Canada inferred from dated sediment cores. Biogeochemistry 40:163–173.

Lorey P, Driscoll CT. 1999. Historical trends of mercury deposition in Adirondack lakes. Environ Sci Technol 33:718–722.

Lucotte M, Montgomery S, Bégin M. 1999b. Mercury dynamics at the flooded soil-water interface in reservoirs of Northern Québec: *in situ* observations. In: Lucotte M, Schetagne R, Thérien N, Langlois C, Tremblay A, editors, Mercury in the biogeochemical cycle. Berlin, Germany: Springer, p. 193–214.

Lucotte M, Montgomery S, Caron B, Kainz M. 1999a. Mercury in natural lakes and unperturbed terrestrial ecosystems of northern Québec. In: Lucotte M, Schetagne R, Thérien N, Langlois C, Tremblay A, editors. Mercury in the biogeochemical cycle. Berlin, Germany: Springer, p. 55–88.

Lucotte M, Mucci A, Hillaire-Marcel C, Pichet P, Grondin A. 1995. Anthropogenic mercury enrichment in remote lakes of northern Québec (Canada). Water Air Soil Pollut 80:467–476.

Lucotte M, Schetagne R, Thérien N, Langlois C, Tremblay A, editors. 1999c. Mercury in the biogeochemical cycle. Berlin, Germany: Springer. 334 p.

Lyons J, Cochran PA, Fago D. 2000. Wisconsin fishes 2000 — status and distribution. Madison (WI): University of Wisconsin Sea Grant Institute. 87 p.

Mahaffey KR. 2000. Recent advances in recognition of low-level methylmercury poisoning. Current Opinion Neurol 13:699–707.

Mahaffey KR, Clickner RP, Bodurow CC. 2004. Blood organic mercury and dietary mercury intake: National Health and Nutrition Examination Survey, 1999 and 2000. Environ Health Perspect 112:562–570.

Martínez-Cortizas A, Pontevedra-Pombal X, Garcia-Rodeja E, Nóvoa-Muñoz JC, Shotyk W. 1999. Mercury in a Spanish peat bog: archive of climate change and atmospheric metal deposition. Science 284:939–942.

Mason RP, Reinfelder JR, Morel FMM. 1996. Uptake, toxicity, and trophic transfer of mercury in a coastal diatom. Environ Sci Technol 30:1835–1845.

Masson S, Tremblay A. 2002. Effects of intensive fishing on the structure of zooplankton communities and mercury levels. Sci Total Environ 304:377–390.

Meili M. 1991. Mercury in boreal forest lake ecosystems. Comprehensive summaries of Uppsala Dissertations from the Faculty of Science, Report No. 336, Uppsala University, Sweden.

Metsaelae T, Rask M. 1989. Mercury concentrations of perch, *Perca fluviatilis* L., in small Finnish headwater lakes with different pH and water colour. Aqua Fennica 19:41–46.

Miskimmin BM, Rudd JWM, Kelly CA. 1992. Influence of dissolved organic carbon, pH, and microbial respiration rates on mercury methylation and demethylation in lake water. Can J Fish Aquat Sci 49:17–22.

Monson BA, Brezonik PL. 1999. Influence of food, aquatic humus, and alkalinity on methylmercury uptake by *Daphnia magna*. Environ Toxicol Chem 18:560–566.

Monteiro LR, Furness RW. 1997. Accelerated increase in mercury contamination in North Atlantic mesopelagic food chains as indicated by time series of seabird feathers. Environ Toxicol Chem 16:2489–2493.

Montgomery S, Lucotte M, Cournoyer L. 2000. The use of stable isotopes to evaluate the importance of fine suspended particulate matter in the transfer of methylmercury to biota in boreal flooded environments. Sci Total Environ 261:33–41.

Morel FMM, Kraepiel AML, Amyot M. 1998. The chemical cycle and bioaccumulation of mercury. Annu Rev Ecol Systematics 29:543–566.

Myers GJ, Davidson PW. 1998. Prenatal methylmercury exposure and children: neurologic, developmental, and behavioral research. Environ Health Perspect 106(Suppl 3):841–847.

NRC 2000. Toxicological effects of methylmercury. National Research Council Committee on the Toxicological Effects of Methylmercury, Washington, DC: National Academies Press. 344 p.

Odin M, Feurtet-Mazel A, Ribeyre F, Boudou A. 1994. Actions and interactions of temperature, pH and photoperiod on mercury bioaccumulation by nymphs of the burrowing mayfly *Hexagenia rigida*, from the sediment contamination source. Environ Toxicol Chem 13:1291–1302.

Parkman H, Meili M. 1993. Mercury in macroinvertebrates from Swedish forest lakes: influence of lake type, habitat, life cycle and food quality. Can J Fish Aquat Sci 50:521–534.

Parks JW, Hamilton AL. 1987. Accelerating recovery of the mercury-contaminated Wabigoon/English River system. Hydrobiologia 149:159–188.

Paterson MJ, Rudd JWM, St Louis VL. 1998. Increases in total and methylmercury in zooplankton following flooding of a peatland reservoir. Environ Sci Technol 32:3868–3874.

Pauly D, Alder J, Bennett E, Christensen V, Tyedmers P, Watson R. 2003. The future for fisheries. Science 302:1359–1361.

Peech Cherewyk K. 2002. Methylmercury bioaccumulation in zooplankton: an assessment of exposure routes and accumulation in newly flooded reservoirs. MS thesis, University of Manitoba, Winnipeg, Canada, 90 p.

Peterson SA, Van Sickle J, Hughes RM, Schacher JA, Echols SF. 2005. A biopsy procedure for determining filet and predicting whole-fish mercury concentration. Arch Environ Contam Toxicol 48:99–107.

Phillips DL. 2001. Mixing models in analyses of diet using multiple stable isotopes: a critique. Oecologia 127:166–170.

Phillips DL, Gregg JW. 2001. Uncertainty in source partitioning using stable isotopes. Oecologia 127:171–179.

Pickhardt PC, Folt CL, Chen CY, Klaue B, Blum JD. 2002. Algal blooms reduce the uptake of toxic methylmercury in freshwater food webs. Proc Nat Acad Sci 99:4419–4423.

Plourde Y, Lucotte M, Pichet P. 1997. Contribution of suspended particulate matter and zooplankton to methylmercury contamination of the food chain in mid-northern Québec (Canada) reservoirs. Can J Fish Aquat Sci 54:821–831.

Porvari P. 1995. Mercury levels of fish in Tucurui hydroelectric reservoir and in River Moju in Amazonia, in the state of Para, Brazil. Sci Total Environ 175:109–117.

Porvari P. 1998. Development of fish mercury concentrations in Finnish reservoirs from 1979 to 1994. Sci Total Environ 213:279–290.

Porvari P, Verta M, Munthe J, Haapanen M. 2003. Forestry practices increase mercury and methyl mercury output from boreal forest catchments. Environ Sci Technol 37:2389–2393.

Rencz AN, O'Driscoll NJ, Hall GEM, Peron T, Telmer K, Burgess NM. 2003. Spatial variation and correlations of mercury levels in the terrestrial and aquatic components of a wetland dominated ecosystem: Kejimkujik Park, Nova Scotia, Canada. Water Air Soil Pollut 143:271–288.

Resh VH, McElvary EP. 1993. Contemporary quantitative approaches to biomonitoring using benthic macroinvertebrates. In: Rosenberg DM, Resh VH, editors. Freshwater biomonitoring and benthic macroinvertebrates. New York (NY): Chapman & Hall, p. 159–194.

Rodgers DW. 1994. You are what you eat and a little bit more: bioenergetics-based models of methylmercury accumulation in fish revisited. In: Watras CJ, Huckabee JW, editors, Mercury pollution: integration and synthesis. Boca Raton (FL): Lewis Publishers, p. 427–439.

Rolfhus KR, Fitzgerald WF. 1995. Linkages between atmospheric mercury deposition and the methylmercury content of marine fish. Water Air Soil Pollut 80:291–297.

Roseman EF, Mills EL, Forney JL, Rudstam LG. 1996. Evaluation of competition between age-0 yellow perch (*Perca flavescens*) and gizzard shad (*Dorosoma cepedianum*) in Oneida Lake, New York. Can J Fish Aquat Sci 53:865–874.

Roulet M, Lucotte M, Canuel R, Farella N, Courcelles M, Guimarães J-R, Mergler D, Amorim M. 2000. Increase in mercury contamination recorded in lacustrine sediments following deforestation in the central Amazon. Chem Geol 165:243–266.

Roulet M, Lucotte M, Farella N, Serique G, Coelho H, Sousa Passos CJ, de Jesus da Silva E, Scavone de Andrade P, Mergler D, Amorim M. 1999. Effects of recent human colonization on the presence of mercury in Amazonian ecosystems. Water Air Soil Pollut 112:297–313.

Roulet M, Lucotte M, Saint-Aubin A, Tran S, Rheault I, Farella N, de Jesus da Silva E, Dezencourt J, Sousa Passos CJ, Santos Soares G, Guimarães JR, Mergler D, Amorim M. 1998. The geochemistry of mercury in central Amazonian soils developed on the Alter-do-Chão formation of the lower Tapajós River valley, Pará state, Brazil. Sci Total Environ 223:1–24.

Rusak JA, Yan ND, Somers KM, Cottingham KL, Micheli F, Carpenter SR, Frost TM, Paterson MJ, McQueen DJ. 2002. Temporal, spatial, and taxonomic patterns of crustacean zooplankton variability in unmanipulated north-temperate lakes. Limnol Oceanogr 47:613–625.

Schetagne R, Verdon R. 1999. Post-impoundment evolution of fish mercury levels at the La Grande complex, Québec, Canada (from 1978 to 1996). In: Lucotte M, Schetagne R, Thérien N, Langlois C, Tremblay A, editors, Mercury in the biogeochemical cycle. Berlin, Germany: Springer, p. 235–258.

Scheuhammer AM, Graham JE. 1999. The bioaccumulation of mercury in aquatic organisms from two similar lakes with differing pH. Ecotoxicology 8:49–56.

Schober SE, Sinks TH, Jones RL, Bolger, PM, McDowell M, Osterloh J, Garrett ES, Canady RA, Dillon CF, Sun Y, Joseph CB, Mahaffey KR. 2003. Blood mercury levels in US children and women of childbearing age, 1999–2000. J Am Med Assoc 289:1667–1674.

Schuster PF, Krabbenhoft DP, Naftz DL, Cecil LD, Olson ML, DeWild JF, Susong DD, Green JR, Abbott ML. 2002. Atmospheric mercury deposition during the last 270 years: a glacial ice core record of natural and anthropogenic sources. Environ Sci Technol 36:2303–2310.

Scott WB, Crossman EJ. 1973. Freshwater fishes of Canada. Ottawa, Ontario: Fish Res Board Can Bull 184. 966 p.

Seda J, Kubecka J. 1997. Long-term biomanipulation of Rimov Reservoir (Czech Republic). Hydrobiologia 345:95–108.

Sellers P, Kelly CA, Rudd JWM, MacHutchon AR. 1996. Photodegradation of methylmercury in lakes. Nature 380:694–697.

Sellers P, Kelly CA, Rudd JWM. 2001. Fluxes of methylmercury to the water column of a drainage lake: the relative importance of internal and external sources. Limnol Oceanogr 46:623–631.

Serdar D, Johnston J, Mueller K, Patrick G. 2001. Mercury concentrations in edible muscle of Lake Whatcom fish. Olympia (WA): Washington State Department of Ecology. Pub No 01-03-012. 28 p. + appendices. http://www.ecy.wa.gov/pubs/0103012.pdf

Shotyk W, Goodsite ME, Roos-Barraclough F, Frei R, Heinemeier J, Asmund G, Lohse C, Hansen TS. 2003. Anthropogenic contributions to atmospheric Hg, Pb and As accumulation recorded by peat cores from southern Greenland and Denmark dated using the ^{14}C "bomb pulse curve." Geochim Cosmochim Acta 67:3991–4011.

Shotyk W, Goodsite ME, Roos-Barraclough F, Givelet N, Le Roux G, Weiss D, Cheburkin AK, Knudsen K, Heinemeier J, van Der Knaap WO, Norton SA, Lohse C. 2005. Accumulation rates and predominant atmospheric sources of natural and anthropogenic Hg and Pb on the Faroe Islands. Geochim Cosmochim Acta 69:1–17.

Simoneau M, Lucotte M, Garceau S, Laliberté D. 2005. Fish growth rates modulate mercury concentrations in walleye (Sander vitreus) from eastern Canadian lakes. Environ Res 98:73–82.

Simonin HA, Gloss SP, Driscoll CT, Schofield CL, Kretser WA, Karcher RW, Symula J. 1994. Mercury in yellow perch from Adirondack drainage lakes (New York, U.S.). In: Watras CJ, Huckabee JW, editors, Mercury pollution: integration and synthesis. Boca Raton (FL): Lewis Publishers, p. 457–469.

Smith AL, Green RH. 1975. Uptake of mercury by freshwater clams (family Unionidae). J Fish Res Board Can 32:1297–1303.

Sonesten L. 2003. Fish mercury levels in lakes — adjusting for Hg and fish-size covariation. Environ Pollut 125:255–265.

Sorensen JA, Kallemeyn LW, Sydor M. 2005. Relationship between mercury accumulation in young-of-the-year yellow perch and water-level fluctuations. Environ Sci Technol 39:9237–9243.

Southerland MT, Stribling JB. 1995. Status of biological criteria development and implementation. In: Davis WS, Simon TP, editors, Biological assessment and criteria: tools for water resource planning and decision making. Boca Raton (FL): Lewis Publishers, p. 81–96.

Southworth GR, Peterson MJ, Ryon MG. 2000. Long-term increased bioaccumulation of mercury in largemouth bass follows reduction of waterborne selenium. Chemosphere 41:1101–1105.

Spry DJ, Wiener JG. 1991. Metal bioavailability and toxicity to fish in low-alkalinity lakes: a critical review. Environ Pollut 71:243–304.

St. Louis VL, Rudd JWM, Kelly CA, Beaty KG, Flett RJ, Roulet NT. 1996. Production and loss of methylmercury and loss of total mercury from boreal forest catchments containing different types of wetlands. Environ Sci Technol 30:2719–2729.

Stafford CP, Haines TA. 2001. Mercury contamination and growth rate in two piscivore populations. Environ Toxicol Chem 20:2099–2101.

Suns K, Hitchin G. 1990. Interrelationships between mercury levels in yearling yellow perch, fish condition and water quality. Water Air Soil Pollut 50:255–265.

Suns K, Hitchin G, Loescher B, Pastorek E, Pearce R. 1987. Metal accumulations in fishes from Muskoka-Haliburton lakes in Ontario (1978–1984). Rexdale, Ontario, Canada: Ontario Ministry of the Environment. Technical Report. 38 p.

Surette C, Lucotte M, Doire J, Tremblay A. 2003. Effects of intensive fishing in three natural lakes of northern Québec, Canada, J Phys IV, Proc 12th Internat Conf on Heavy Metals in the Environment 107:1443.

Surette C, Lucotte M, Tremblay A. 2006. Influence of intensive fishing on the partitioning of mercury and methylmercury in three lakes of Northern Québec. Sci Total Environ 368(1):248–261.

Svobodova Z, Dusek L, Hejtmanek M, Vykusova B, Smid R. 1999. Bioaccumulation of mercury in various fish species from Orlik and Kamyk Water Reservoirs in the Czech Republic. Ecotoxicol Environ Safety 43:231–240.

Swain EB, Engstrom DR, Brigham ME, Henning TA, Brezonik PL. 1992. Increasing rates of atmospheric mercury deposition in midcontinental North America. Science 257:784–787.

Thomann RV, Mahoney JD, Mueller R. 1995. Steady-state model of biota sediment accumulation factor for metals in two marine bivalves. Environ Toxicol Chem 14:1989–1998.

Thorpe J. 1977. Synopsis of biological data on the perch *Perca fluviatilis* (Linnaeus, 1758) and *Perca flavescens* (Mitchill, 1814). Rome, Italy: Food and Agriculture Organization of the United Nations. FAO Fisheries Synopsis No 113. 138 p.

Tremblay G, Legendre P, Doyon J-F, Verdon R, Schetagne R. 1998. The use of polynomial regression analysis with indicator variables for interpretation of mercury in fish data. Biogeochemistry 40:189–201.

Tremblay A, Lucotte M. 1997. Accumulation of total mercury and methylmercury in insect larvae of hydroelectric-reservoirs. Can J Fish Aquat Sci 54:832–841.

Tsui MTK, Wang WX. 2004. Uptake and elimination routes of inorganic mercury and methylmercury in *Daphnia magna*. Environ Sci Technol 38:808–816.

[USEPA] US Environmental Protection Agency. 2000. Guidance for assessing chemical contamination data for use in fish advisories. Vol. 1: Fish sampling and analysis, 3rd ed. Washington, DC: USEPA Office of Water. EPA 823-B-00-007. http://www.epa.gov/waterscience/fishadvice/volume1/index.html

[USEPA] US Environmental Protection Agency. 2001. Update: national listing of fish and wildlife advisories. Washington, DC: USEPA, Office of Water. Fact Sheet EPA-823-F-01-010.

[USEPA] US Environmental Protection Agency. 2005. 2004 National listing of fish advisories. Washington, DC: USEPA, Office of Water. Fact Sheet EPA-823-F-05-004.

Vander Zanden MJ, Vadeboncoeur Y. 2002. Fishes as integrators of benthic and pelagic food webs in lakes. Ecology 83:2152–2161.

Verdon R, Tremblay A. 1999. Mercury accumulation in fish from the La Grande complex: influence of feeding habits and concentrations of mercury in ingested prey. In: Lucotte M, Schetagne R, Thérien N, Langlois C, Tremblay A, editors, Mercury in the biogeochemical cycle. Berlin, Germany: Springer, p. 215–233.

Verta M. 1990. Changes in fish mercury concentrations in an intensively fished lake. Can J Fish Aquat Sci 47:1888–1897.

Wang XF, Fisher NS. 1999. Assimilation efficiencies of chemical contaminants in aquatic invertebrates: a synthesis. Environ Toxicol Chem 18:2034–2045.

Watras CJ, Back RC, Halvorsen S, Hudson RJM, Morrison KA, Wente SP. 1998. Bioaccumulation of mercury in pelagic freshwater food webs. Sci Total Environ 219:183–208.

Watras CJ, Bloom NS, Hudson RJM, Gherini S, Munson R, Claas SA, Morrison KA, Hurley J, Wiener JG, Fitzgerald WF, Mason R, Vandal G, Powell D, Rada R, Rislove L, Winfrey M, Elder J, Krabbenhoft D, Andren AW, Babiarz C, Porcella DB, Huckabee JW. 1994. Sources and fates of mercury and methylmercury in Wisconsin lakes. In: Watras CJ, Huckabee JW, editors, Mercury pollution: integration and synthesis. Boca Raton (FL): Lewis Publishers, p. 153–177.

Watras CJ, Morrison KA, Kratz TK. 2002. Seasonal enrichment and depletion of Hg and SO_4 in Little Rock Lake: relationship to seasonal changes in atmospheric deposition. Can J Fish Aquat Sci 59:1660–1667.

Weigel BM, Wang LZ, Rasmussen PW, Butcher JT, Stewart PM, Simon TP, Wiley MJ. 2003. Relative influence of variables at multiple spatial scales on stream macroinvertebrates in the Northern Lakes and Forest ecoregion, USA. Freshwat Biol 48:1440–1461.

Wiener JG, Fitzgerald WF, Watras CJ, Rada RG. 1990. Partitioning and bioavailability of mercury in an experimentally acidified Wisconsin lake. Environ Toxicol Chem 9:909–918.

Wiener JG, Krabbenhoft DP, Heinz GH, Scheuhammer AM. 2003. Ecotoxicology of mercury. In: Hoffman DJ, Rattner BA, Burton GA Jr, Cairns J Jr, editors, Handbook of ecotoxicology, 2nd ed. Boca Raton (FL): CRC Press, p. 409–463.

Wiener JG, Shields PJ. 2000. Mercury in the Sudbury River (Massachusetts, USA): pollution history and a synthesis of recent research. Can J Fish Aquat Sci 57:1053–1061.

Wolfe MF, Schwarzbach S, Sulaiman RA. 1998. Effects of mercury on wildlife: a comprehensive review. Environ Toxicol Chem 17:146–160.

5 Wildlife Indicators

Marti F. Wolfe, Thomas Atkeson, William Bowerman, Joanna Burger, David C. Evers, Michael W. Murray, and Edward Zillioux

ABSTRACT

A number of wildlife species are potentially at greater risk of elevated mercury exposures, and development of a monitoring network for mercury in wildlife must take into account numerous variables that can affect exposures (and potentially effects). Because they are generally at the receiving end of the mercury cycle (following releases of inorganic mercury, atmospheric and aquatic cycling and bioaccumulation), numerous factors upstream can affect the amount of mercury available for uptake. As is the case with aquatic biota, methylmercury is of particular concern due to its ability to accumulate to greater extents in wildlife. A number of factors can affect methylmercury uptake in wildlife, including diet (including seasonal or inter-annual variations) and functional niche, location (including consideration of exposure differences for migratory species), age, sex, reproductive status, nutritive status, and disease incidence. In identifying potentially good indicator species for mercury exposure, desirable characteristics include a well-described life history, relatively common and widespread distribution, capacity to accumulate mercury in a predictable fashion (including sensitive to changes in mercury levels, and ideally occurring across a gradient of contaminant levels), easily sampled and adequate population size, and having data on natural physiological variability. Sample collection for mercury analysis must consider methodological factors such as live (e.g., feathers, hair/fur, blood) vs. dead (e.g. internal organs) specimens, time of exposure in relation to tissue sampled (e.g. more recent exposures in blood or eggs vs. longer-term exposures in kidney, fur, or feathers), site of the collection within tissue, potential for and extent of detoxification/depuration, differences within clutches, feathers, or hair locations in birds, and potential for exogenous contamination. In addition to consideration of mercury exposures in developing a monitoring network, effects of mercury could also be considered, including assessments across several levels of biological organization. While several endpoints of mercury toxicity have been identified in wildlife (including growth, reproduction, and neurological), solid biomarkers of mercury effect meeting desirable criteria have to date not been identified. Based on research to date on numerous wildlife species and consideration of indicator criteria identified here, candidate wildlife species for bioindicators of mercury exposure, by habitat type, include the following:

terrestrial — Bicknell's thrush and raccoon; lake — common loon; freshwater wetland — tree swallow; lake/coastal — herring gull, bald eagle and common tern; riverine — mink; estuarine — saltmarsh sharp-tailed and seaside sparrows; nearshore marine — harbor porpoise; offshore marine — Leach's storm petrel; comparison across aquatic habitats — belted kingfisher. It is recommended that monitoring be done annually, considering time after arrival at breeding site for migratory species. Several medium- to long-term monitoring efforts have been conducted for mercury in wildlife (including for egrets and herring gulls). However, clear consideration of the numerous factors affecting mercury uptake and mobilization within individuals, intra- and interspecies variability, and resulting statistical issues must be taken into account in designing a monitoring network that can adequately address questions on spatial and temporal trends of mercury exposure (and potentially effects) in wildlife.

5.1 INTRODUCTION

A bioindicator can be defined as an organism (biological unit or derivative) that responds predictably to contamination in ways that are readily observable and quantifiable (Zillioux and Newman 2003). This response could be at any level of physiological or ecosystem organization from molecular or cellular at 1 end of the spectrum to population or community at the other end. Wildlife species are good indicators of the status of contaminants in the environment because they reflect not just the presence, but also the bioavailability of the contaminant of interest; integrate over time and space and among local, regional, and global sources; and respond to toxic insult in ways that are relevant to human health at both the whole organism and sub-organismal levels.

The effects of mercury in wildlife species are well established and have been the subject of several reviews (Scheuhammer 1987; Scheuhammer 1990; Zillioux et al. 1993; Heinz 1996; Thompson 1996; Burger and Gochfeld 1997; Wolfe et al. 1998; Eisler 2006).

5.1.1 Objectives

Several candidate wildlife indicators are suggested and discussed in this chapter. In addition, we recognize that valuable sources of data on residue-effect relationships are available to assist in the selection of habitat-specific indicators (Jarvinen and Ankley 1999; USCOE and USEPA 2005). Although this chapter emphasizes animals, similar considerations and literature exist for plants and microorganisms as bioindicators and biomarkers (National Research Council 1989; USEPA 1997; Gawel et al. 2001; Citterio et al. 2002; Yuska et al. 2003).

In choosing wildlife indicators of mercury contamination, emphasis should go to 3 key considerations: 1) efficacy in quantifying the probability that mercury in the environment will produce an adverse effect in exposed organisms or populations; 2) the degree of harm that may be anticipated; 3) and the integration of these data to characterize environmental health.

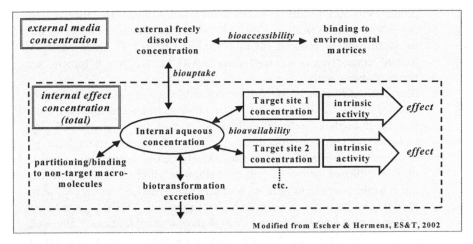

FIGURE 5.1 Pathways of bioaccessibility, biouptake, and bioavailability leading to exposure. (*Source:* Modified from Escher and Hermens 2002.)

An additional consideration is the species' usefulness as biomonitors of trends in mercury loading on their ecosystem. In any case, the value of a well-selected bioindicator lies in its ability to integrate all the complex processes leading to the adverse consequence. Figure 5.1 traces schematically the process pathways of bio-accessibility, biouptake, and bioavailability (as defined below) that must be complete before reaching the target-organ dose at which harm might be caused to humans or wildlife. Such pathways of exposure are typically habitat- and organism-specific.

The terms "bioaccessibility," "biouptake ," and "bioavailability," as used in this chapter, are defined below in the context of 2 primary considerations: 1) the major and best-characterized route of exposure of wildlife to environmental mercury con-tamination is through the aquatic food web; and 2) mercury incorporated into fish, piscivorous wildlife and their higher predators is predominately (generally >95%) in the form of methylmercury (MeHg). Although we are concerned here primarily with aquatic systems, it must be noted that very recent work has identified an entirely terrestrial pathway by which vertebrates are exposed to MeHg; this research is in its infancy but should be followed closely, as the mechanisms by which MeHg is transferred in nonaquatic systems are poorly understood (Rimmer et al. 2005).

- *Bioaccessibility: the conversion of mercuric mercury (Hg (II)) to methyl-mercury (CH_3Hg^+ or MeHg) in an environment accessible to organisms at the base of the aquatic food web.* This is the most critical step in the delivery of environmental mercury to target organs/molecules in fish and other wildlife species. The formation of MeHg, the principal environmen-tally toxic species, is necessary for accessibility of Hg to the aquatic food web and sets the stage for the biological uptake. MeHg is the main product of the natural biomethylation reaction carried out by sulfate-reducing bacteria principally at or near the sediment/water interface. Hg (II) is the primary substrate for the biomethylation reaction. Biomethylation is rate-dependent upon a variety of biogeochemical conditions conducive for this transformation to proceed (as discussed in Chapter 3). Once MeHg is formed, connection of the contaminant with the receptor completes the bioaccessibility step.

- *Biouptake: diffusion of MeHg through a biological membrane into the internal cellular and plasma environment of an organism.* This diffusion may be through an external cellular membrane (as in single-cell phytoplankton or simple multicellular infaunal organisms), or through the gut or caecum epithelia of prey species. MeHg can also enter an organism through diffusion across the gill epithelium although, in the case of fish, this is a minor source of uptake given the comparatively low concentration of MeHg in the water column. In higher organisms, MeHg ingested from prey species is readily absorbed through the intestinal mucosa. Embryonic uptake of MeHg occurs by absorption from stored food in the egg of oviparous and ovoviviparous species, and by diffusion across the placental "barrier" in mammals.

- *Bioavailability: the delivery of MeHg to a target organ or site of toxic action.* Once taken up, MeHg is highly mobile and distributed throughout the body. Nevertheless, not all MeHg that enters the body is actually bioavailable. Several natural elimination and detoxification processes remove MeHg from the circulatory system before delivery to target organs/molecules (principally those of the central nervous system). Examples are removal from the systemic circulatory system through accumulation in hair and feathers, and presystemic elimination by metabolic transformation to Hg (II) in the liver and subsequent excretion in feces. MeHg also accumulates in non-target tissues, such as muscle and kidney, in each of which MeHg has its own biological half-life.

Wildlife indicators can establish baseline conditions, act as early warning signals of environmental problems, identify the extent of contamination, define critical pathways and responses at multiple trophic levels, as well as integrate biological exposure with the physical and chemical environment (Farrington 1991). Indicator selection is based on a combination of criteria or characteristics that include (Jenkins 1981):

- Well-characterized life history
- Capable of concentrating and accumulating contaminant(s) of concern
- Common in the environment
- Geographically widespread
- Sensitive and hence indicative of change
- Easily collected and measured
- Adequate size to permit resampling of tissue
- Occurrence in both polluted and unpolluted areas
- Display correlation with environmental levels of contaminants
- Has background data on the natural condition

Burger and Gochfeld (2000c, 2004) list key features of a biomonitoring plan that fulfill requirements of biological, methodological, or societal relevance. These attributes are further discussed in Sections 5.4 and 5.8.

Wildlife indicators of mercury exposure and trends are important elements of a comprehensive approach to assess mercury in the environment and the monitoring of trends that may assist regulators and the regulated community in long-term evaluation of the need and usefulness of mercury source controls. It is important to understand, however, that bioindicator data alone are insufficient to answer such critical questions as identification of mercury sources, or the relative importance of local, regional, and global inputs of mercury sources to atmospheric deposition and environmental loading in specific areas.

5.2 ISSUES OF CONCERN

5.2.1 GEOGRAPHICAL AND HABITAT DIFFERENCES

Geography and habitat variability affect MeHg production, bioaccessibility, and uptake into wildlife. Interpretation of mercury in wildlife also requires a working knowledge of sex, age, and tissue differences (Evers et al. 2005). Biogeochemical differences in aquatic and terrestrial systems are particularly important determinants of Hg methylation, as discussed in previous chapters for water and fish.

Continental Hg patterns are therefore dictated by large-scale atmospheric deposition patterns, point source emissions (and effluents), and ecosystem processes. Using a standard indicator species, Evers et al. (1998, 2003) documented an increasing west-east pattern in continental MeHg concentrations in blood and eggs for the Common Loon (*Gavia immer*) (Figure 5.2). Although many areas exist throughout North America where Hg deposition probably poses risk to biota, general west-east weather patterns do appear to influence overall MeHg bioavailability and contribute to the well-known "tail-pipe" condition of northeastern North America. Documented aquatic systems outside of the Northeast where MeHg concentration is elevated and, at least in part, related to atmospheric deposition are north-central Wisconsin and the western Upper Peninsula of Michigan (primarily because of high acidic lake systems) (Meyer et al. 1998; Fevold et al. 2003) and southernmost Florida (Frederick et al. 2002; Frederick et al. 2004). Vast and highly acidic aquatic systems in eastern Ontario and western Quebec also remain as troublesome areas for elevated risk of Hg to high trophic level piscivores because of continued acidic conditions related to anthropogenic input of sulfur dioxide (Doka et al. 2003). Mercury deposition in the West presents some unique considerations. Throughout the West as a region, mercury inputs from legacy mining greatly exceed inputs from atmospheric deposition, but where coal-fired electric power generation is used, very localized atmospheric Hg concentrations sometimes exceed even those found in the highly urbanized East. For the 3 coastal western states, trans-Pacific transport of atmospheric Hg from Asian sources is a recent and increasing input. The importance of this contribution to total Hg loading in the coastal states is currently under examination (Fitzgerald and Mason 1997; Weiss-Penzias et al. 2003; Seigneur et al. 2004; Jaffe et al. 2005).

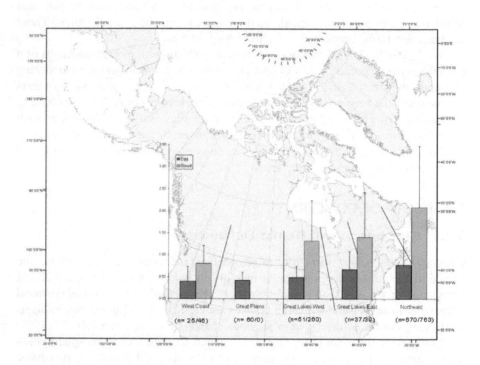

FIGURE 5.2 Continental cross-section of MeHg bioavailability in common loon blood and eggs. Mercury concentrations are arithmetic means and associated 1 SD in ppm, ww. Sample size in parentheses are first eggs and then blood. (*Source:* From Evers et al. 1998, 2003b.)

We have categorized 4 major habitat types: 1) marine, 2) estuarine, 3) freshwater, and 4) terrestrial. Differences in mercury cycling among the major habitat types are not well understood, although most studies characterizing biotic uptake of Hg through complete food chains have focused on freshwater environs. There are more data on Hg in marine mammals than in freshwater mammals, but the movement of Hg through all trophic levels in marine food chains is poorly known. Marine systems and their respective indicators reflect forage guilds that use the shoreline as well as nearshore and offshore habitats. Some research on Hg exposure in birds foraging in coastal and pelagic habitats within the Canadian Maritimes indicates spatial variation that may be related to forage base among other factors (Burgess, N., personal communication).

A handful of studies have compared species Hg levels across different habitat types. Welch (1994) found juvenile bald eagle blood Hg levels were significantly higher in freshwater versus marine systems. Studies using belted kingfishers across all 4 habitats documented similar patterns; blood Hg levels significantly increased from marine to estuarine to riverine to lakes (Evers et al. 2005). The biogeochemical factors that influence Hg methylation and bioavailability within each of these major habitat categories are described in Chapters 2 and 3 and indicate that freshwater aquatic systems associated with wetlands and acidic environments are at greatest risk.

Regional differences in hydrology such as flow patterns, rates, and periodicity, as well as dry-down and rewetting in some environments, may occur seasonally or as a consequence of water management strategies. Watershed drainage and flow rates affect Hg transport and residence times, and nutrient and sulfate loading which, in turn, influence Hg methylation and bioaccessibility. Periodic dry-downs and rewetting affect the sulfur cycle through sulfide oxidation and sulfate reduction, respectively. In turn, Hg methylation by sulfate-reducing bacteria is the probable cause of large spikes in available MeHg in these areas during and immediately following periods of rewetting (Krabbenhoft et al. 1998). Biota Hg is generally higher in reservoirs, particularly new reservoirs, than in other areas of contiguous watersheds. This "new reservoir effect" typically diminishes with time but the rate of change is strongly influenced by latitudinal factors; elevated biota Hg levels may persist for many years in higher latitude reservoirs (Bodaly et al. 1984) while the effect may be fleeting or undetectable in lower latitudes (Abernathy and Cumbie 1977). Older reservoirs, particularly those with bathymetry that serve as large areas of suitable habitat for bacteria to methylate Hg, are potential high-risk scenarios. Such reservoirs in northern New England that have high organic content shorelines and slow water drawdowns through summer and fall (e.g., water storage reservoirs) are documented with greatly elevated biotic Hg levels (Evers and Reaman 1997).

Habitat differences also influence trophic structure, with the length of food chains affecting the degree of bioaccumulation of Hg in top predators. Prey species availability in different habitats may strongly influence accumulation of Hg in predators. Porcella et al. (2004) reviewed raccoon dietary composition and showed that, among food groups dominating raccoon foraging under various conditions, progressively lower dietary Hg is available when habitat or seasonal foraging opportunities are restricted to lower levels in the food chain. The Florida Panther (*Puma concolor coryi*) has been shown to accumulate high levels of tissue Hg when feeding on raccoons in the central Everglades, whereas in the nearby Fakahatchee Strand, where their normal diet of deer and wild hog is available, panthers accumulate much lower levels of Hg (Roelke et al. 1991). Ecosystem nutrient status also influences the bioaccessibility of mercury to higher trophic levels. Eutrophication resulting in the proliferation of lower trophic levels can cause a "biodilution effect" that effectively limits mercury available to predator species (Chen et al. 2000; Stafford and Haines 2001). On the other hand, poor nutrient status among individual species may compromise the ability of affected species to process and detoxify dietary Hg. Differences in the form and concentration of environmental selenium may also affect Hg detoxification mechanisms in some species. In marine mammals, for example, frequently observed molar ratios of liver Hg to Se of 1:1 suggest that this highly insoluble form (i.e., mercuric selenide) sequesters Hg and prevents further toxicity (Wagemann et al. 2000), but also see Caurant et al. (1996) for limits to this process.

In the case of marine mammals, geographic and habitat differences — even for individuals — can be quite diverse. Some species may have distinctly separated (via migration routes) foraging and breeding habitats (e.g., for the California gray whale (*Eschrichtius robustus*) or minke whale (*Balaenoptera acutorostrata*)), while others are largely nonmigratory (e.g., some pelagic dolphins and harbor seals (*Phoca vitulina*)). Even some species that are not migratory move to new foraging locations based on prey availability (e.g., long-finned pilot whales (*Globicephala melas*)), and others range widely and may switch to different foraging dive depths at different times of year (e.g., hooded seals (*Cystophora cristata*) in the North Atlantic) (Bjorge 2001). Mercury loadings to marine systems will vary; in addition to an assumed more broadly uniform pattern of air deposition across wide areas, recent research has highlighted the potential for increased deposition in high latitude regions during polar springtimes (see Chapter 2), as well as the potential for some freshwater drainages to contribute significant loadings. Some studies have revealed spatial trends in mercury levels in marine mammals, with for example higher levels in St. Lawrence beluga whales (*Delphinapterus leucas*) than Arctic belugas, and higher mercury levels in muscle, kidney, and liver tissues in belugas in the Western as compared to Eastern Arctic (Wagemann et al. 1996).

5.2.2 METHODOLOGICAL ISSUES

Both the development and application of bioindicators present a number of methodological considerations. One key requirement is to relate dose/effects studies in the laboratory, and residue levels/effects studies in the field. For many years, these studies were conducted by different groups of scientists, and the connections were not made (Eisler 1987). Ideally, we should use bioindicators where there are clear links between exposure levels, tissue levels, and effects (Burger and Gochfeld 2003). The most useful bioindicators of those we suggest are those where the connections have been clearly made.

A knowledge of physiology and pharmacokinetics is needed (Farris et al. 1993; Monteiro and Furness 2001). Levels of mercury normally vary among internal tissues, and the time to equilibrate within each tissue varies. For example, blood mercury levels normally reflect very recent exposure, while brain and liver levels reflect longer-term exposure. Tissue-specific mechanisms of detoxification and sequestration, among other processes, must be understood to define the bioactive moiety in observed tissue burdens before a clear expression of toxicity can be derived (Wood et al. 1997).

Several factors must be considered when collecting samples, and in reporting results of residue analysis: sample collection location, whether the samples were taken from live versus dead specimens, how representative the sample residue is of internal mercury levels, including consideration of sampling location within organs; possible differences within and between clutches, locations (on the animal) from which feathers or hair samples were taken, and potential for exogenous contamination.

For threatened or endangered species, or species of special concern, it is often necessary to analyze specimens that have died of causes not directly attributable to mercury. Bird eggs that have been abandoned or flooded out may be used for analyses. However, if the eggs were pushed out of the nest by parents that are incubating the rest of the clutch, the reason for rejection of the egg must be considered in order to properly interpret mercury residue levels. Similarly, birds killed by predators may be suitable for analysis, but the internal tissues of sick or emaciated birds should not be used for residue analysis because in some studies, error has resulted from remobilization of mercury (Ensor et al. 1992; Sundlof et al. 1994).

However, investigations that excluded emaciated birds indicate that comparison of mercury concentrations between live and dead specimens may be useful. The direction of error is not always the same, and in some cases, live birds have higher levels (Burger 1995).

The specific site of tissue collection may affect residue levels significantly. For example, samples from the anterior portions of fish can have significantly higher levels of mercury than posterior sections (Cuvin-Aralar and Furness 1990; Furness et al. 1990; Allen 1994; Yediler and Jacobs 1995). Different parts of the liver can accumulate different levels of mercury; because liver Hg and MeHg do not concentrate at a proportionate rate, care in interpretation of liver Hg levels is needed (Scheuhammer et al. 1998b). Caution should also be used when examining mercury levels in eggs because mercury is often higher in the first-laid egg and lowest in the last-laid egg. Therefore, within-clutch differences in egg mercury levels can be significant and knowledge of egg-laying order is needed to minimize variation in interpretation (Becker 1992). Evers et al. (2003b) found an average within-clutch difference of mercury levels in common loon eggs of 25%. Feather mercury levels follow a similar pattern. Within a molt, either body or remigial, the first-grown feathers are higher in mercury than the last-grown feathers (as long as the diet does not change during the molt) (Burger 1993). In addition, depending on molt patterns, different feathers may represent mercury uptake in different geographic areas (Furness et al. 1986; Thompson et al. 1992; Burger 1993; Bowerman et al. 1994). Some birds, such as loons, have full remigial molts and therefore choice of flight feathers is not as critical (Evers et al. 1998). Bowerman et al. (1994) found no significant differences among feather type collected (body, primary, secondary, tail) for Hg within a bald eagle breeding area, and thus concluded that the feather type is not critical for eagles because they typically exhibit a full body and remigial molt in the spring. These variant findings reinforce the importance of carefully considering species differences, tissue types, and collection methods.

5.3 HOST FACTORS

The ecological constraints of any species that is a candidate for monitoring environmental contaminants must be well characterized. Diet, functional niche, migratory status, and home range size influence a species' suitability as an indicator. Seasonal changes in these parameters also will be reflected in contaminant concentrations. An animal's age and sex overall body condition and health status also influence its suitability as indicator (Evers et al. 2005). All of these factors can also alter the bioavailability, toxicokinetics and toxicodynamics of a contaminant, thereby altering uptake, distribution, and effects. Whole body retention of mercury was greater in females than males in 3 mouse strains tested (Nielsen et al. 1994). Lactating pilot whales were less able to demethylate mercury by forming Hg-Se complexes, indicating greater MeHg transference to the nursing calves (Caurant et al. 1996). Possible co-exposure to other environmental contaminants that may modify the organism's response to mercury is also important to determine (Batel et al. 1993; Moore et al. 1999; Mason et al. 2000; Newland and Paletz 2000; Seegal and Bemis

2000; Shipp et al. 2000; Burger 2002; Lee and Yang 2002; Wayland et al. 2002; Wayland et al. 2003).

5.3.1 BIOAVAILABILITY

Ingested Hg may be either inorganic or organic, although, as noted previously, MeHg predominates in higher trophic level organisms. Most inorganic mercury in the environment is in the more thermodynamically stable divalent (mercuric) form. Methylmercury is readily absorbed from the gastrointestinal tract (90 to 95%), whereas inorganic salts of Hg are less readily absorbed (7 to 15%). In the liver, Hg binds to glutathione, cysteine, and other sulfhydryl-containing ligands. These complexes are secreted in the bile, releasing the Hg for reabsorption from the gut (Doi 1991). Demethylation also occurs in the liver, thus reducing toxicity and reabsorption potentials (Komsta-Szumska et al. 1983; Farris et al. 1993; Nordenhall et al. 1998). In blood, MeHg distributes 90% to red blood cells, and 10% to plasma. Inorganic Hg distributes approximately evenly or with a cell:plasma ratio of ≥ 2 (Aihara and Sharma 1986). O'Connor and Nielsen (1981) found that length of exposure was a better predictor of tissue residue level than dose in otters, but that higher doses produced an earlier onset of clinical signs.

5.3.2 TOXICOKINETICS AND TOXICODYNAMICS

Methylmercury readily crosses the blood-brain barrier, whereas inorganic Hg does so poorly. The transport of MeHg into the brain is mediated by its affinity for the anionic form of sulfhydryl groups. This led Aschner (Aschner and Aschner 1999; Aschner 1990) to propose a mechanism of "molecular mimicry" in which the carrier was an amino acid. Transport of MeHg across the blood-brain barrier in the rat as MeHg–L-cysteine complex has since been described (Kerper et al. 1992). Demethylation occurs in brain tissue, as evidenced by the observation that the longer the time period between exposure to MeHg and measurement of brain tissue residue, the greater the proportion of inorganic mercury (Norseth and Clarkson 1970; Lind et al. 1988; Davis et al. 1994). MeHg is also converted to mercuric Hg in other tissues, but the rate of demethylation varies both with tissue (Dock et al. 1994; Wagemann et al. 1998; Pingree et al. 2001) and among species for a given tissue (Omata et al. 1986, 1988).

Both inorganic and organic Hg are excreted primarily in feces; 98 days after administration of a radio-labeled dose of MeHg to rats, 65% of the dose was recovered in the feces as inorganic mercury, and 15% as organic mercury. Urinary excretion accounted for less than 5% of the dose, although urinary excretion of inorganic Hg increased with increasing time after exposure. Fur or hair is also an important route of excretion for both methyl and inorganic Hg. On an average of species and tissues, the biological half-life of MeHg in mammals is about 70 days; for inorganic Hg about 40 days (Farris and Dedrick 1993). The half-life of Hg in non-molting seabirds has been estimated as 60 days (Monteiro and Furness 1995); in comparison, the half-life of MeHg in blood of common loon chicks undergoing feather molt is 3 days (Fournier et al. 2002).

5.4 TYPES OF BIOINDICATORS

5.4.1 INDICATORS OF EXPOSURE

Mercury exposure can be measured in many compartments in an organism. Objectives and logistical considerations dictate compartment choice. Nonlethal choices include abandoned eggs, blood, feathers, fur, scales; and for lizards and snakes, tails. With some additional effort muscle and even organ biopsies may be included in this category. These compartments provide avenues for sampling individuals over time and can provide short- and long-term insights on mercury bioaccumulation. Availability of fresh carcasses can also provide information on mercury exposure; emphasis is typically on the liver, kidney, spleen, muscle, and brain.

5.4.2 INDICATORS OF EFFECT

Multiple levels of biological organization should be investigated when determining mercury effects and should include molecular, cellular, individual, population, and ideally, community levels. These efforts can be further organized into cause-and-effect, correlative, and weight of evidence. Our ability to use these approaches is generally related to feasibility of laboratory or mesocosm experiments and *in situ* studies. Molecular ecology and epidemiology, particularly the replicability of genetic analysis, provide increasing ability to examine effects of mercury (see Section 5.7). Investigations into the impacts of mercury on individuals can be categorized into physiological/functional, morphological, behavioral, reproductive, and demographic. Useful endpoints include those that affect growth, viability, reproductive or developmental success, including behavior, immunological effects, neurological impairment and neurohistological lesions, and teratology.

Compared to organic contaminants and their documented morphological impacts to individuals in eagles and cormorants (Welch 1994; Grasman et al. 1998) and eggs (e.g., eggshell thinning) (Fox et al. 1980; Mineau et al. 1984; Risebrough 1986; Gilbertson et al. 1991; Fox 1992), mercury impacts are primarily based on neurological damage. Among wildlife species, impaired behavior related to mercury exposure has been documented in common loons (Nocera and Taylor 1998; Counard 2000; Olsen et al. 2000; Evers et al. 2004), mallards (Heinz 1975), quail (Thaxton and Parkhurst 1973), fish (Hilmy et al. 1987; Webber and Haines 2003), frogs (Britson and Threlkeld 1998), as well as in humans and laboratory animals (Finocchio et al. 1980; Bornhausen and Hagen 1984; Grandjean et al. 1997; Houpt et al. 1988; Grandjean et al. 1998; Kim et al. 2000). Reproductive anomalies related to mercury have been documented in laboratory studies (Fimreite 1971; Fimreite and Karstad 1971; Heinz 1976; Heinz 1979), as well as in the wild. Field studies represent areas impacted by waterborne point sources (e.g., industrial sites, chlor-alkali plants (Fimreite et al. 1970; Fimreite et al. 1971; Gilbertson 1974; Barr 1986), mines (Wolfe and Norman 1998b; Russell 2003), airborne point sources (Evers and Jodice 2002; Florida Department of Environmental Protection 2003), and more remote systems largely driven by atmospheric deposition from regional and potentially global sources (Fitzgerald and Mason 1996; Burgess et al. 1998; Evers et al. 1998).

Studies of effects on common loon populations indicate significant reductions in reproductive success for some high-risk populations (Burgess et al. 1998; Meyer et al. 1998; Evers et al. 2004), which are related to smaller egg size (Evers et al. 2003b), reduced incubation effort (thus lower hatchability), and lower chick survival. Adult survivorship measures of mercury effects are also difficult endpoints to measure but are important because of the ability for mercury to bioaccumulate (i.e., input is greater than output that includes demethylation, sequestering, and depuration). Long-lived, high trophic level species are likely at greatest risk. High-risk common loon males (i.e., blood mercury levels >3.0 ppm, ww) have mean annual accumulation rates of more than 9% (Evers et al. 1998).

5.5 CANDIDATE BIOINDICATOR SPECIES

5.5.1 MAMMALS

5.5.1.1 Mink (*Mustela vison*)

Mink are widely distributed across North America aquatic habitats. Although mink are prey generalists, they primarily feed on aquatic organisms (depending on geography, habitat use, and season). Their home range varies from 8 ha to over 760 ha, with males moving vastly greater differences than females (Baker 1983). Mink have been identified as being particularly sensitive to environmental mercury levels and, because of the availability of trapper-oriented carcasses, exposure levels are relatively well known across large geographic areas of North America (Wobeser et al. 1976; Kucera 1983; Wren 1986; Wren et al. 1986; Foley et al. 1988; Evans et al. 1998; Mierle et al. 2000; Yates et al. 2005). Field efforts generally rely on organ tissues such as liver, kidney, and brain, but fur and muscle are also collected. The mink is a strong indicator species because of large existing databases, laboratory dosing studies (Aulerich et al. 1973; Aulerich et al. 1974; Wobeser et al. 1976; Wren et al. 1987a, 1987b; Dansereau et al. 1999; Basu et al. 2003b; Major et al. 2005; Yates et al. 2005), widespread range, and relatively ubiquitous aquatic habitat use (Yates et al. 2005).

5.5.1.2 River Otter (*Lontra canadensis*)

River otter are primarily piscivores although crayfish and mussels are also important prey items. Reintroduction programs have assisted in a recolonization of much of their former North American range. Because of their large home range (up to 177 km^2) and ability for long-distance movements (up to 160 km) (Baker 1983), body burdens of mercury in otter are generally not reflective of a specific water body. Adult females without young have the smallest home ranges, whereas young males have the greatest potential for long-distance dispersal. Otter are commonly used as an indicator of aquatic mercury levels (Cumbie 1975b; Kucera 1983; Wren 1984; Wren et al. 1986; Foley et al. 1988; Evans 1995; Evans et al. 1998; Mierle et al. 2000; Wren 1984; Yates et al. 2005). Field measurements of mercury exposure are generally based on tissues similar to mink. Unlike mink, however, laboratory information on Hg effects in otter is limited to 1 feeding study (O'Connor and Nielsen 1981). Although the otter is approximately 10 times the weight of mink, and therefore will tend to forage on larger prey items, mercury levels in each species from the same area are generally similar or even higher in mink (Yates et al. 2005).

5.5.1.3 Raccoon (Procyon lotor)

The raccoon is widely distributed across most forested areas of North America. In raccoon studies where a large number of samples were collected, hair Hg correlated well with other tissues (e.g., Cumbie 1975a; Wolfe and Norman 1998; Lord et al. 2002). Hair mercury analysis for raccoons, therefore, reflects accumulation levels in plant and animal food consumed, which varies with seasonal availability. For example, in the Florida Everglades area of maximum Hg sediment levels, maximum total Hg concentrations in potential prey species based on wet weight were 57.8 ng/g in aquatic vegetation, 74.4 ng/g in segmented worms, 496 ng/g in aquatic insects, and 1160 ng/g in fish (Loftus et al. 1998). Apple snails in this area averaged 67 ng/g (Eisemann et al. 1997) and crayfish ranged from 32 ng/g in tissue to 81 ng/g in exoskeleton (data from DG Rumbold as cited in Porcella et al. 2004). However, in a review of mercury bioaccumulation in benthic invertebrates, Pennuto et al. (2005) noted that mercury sorbed to exoskeleton is not likely to be bioavailable to predators. Raccoons can be particularly valuable to define long-term trends in food-chain proliferation if sampling is conducted during the same seasonal period every year that is associated with the maximum mercury incorporation into hair tissue. Although this may reflect the greatest biouptake from prey species, hair incorporation of MeHg may lag behind the critical foraging period. The optimum sampling window also will be constrained by the hair biological half-life ($BT_{1/2}$), estimated to be about 130 days. Additional approaches to assess mercury uptake in raccoons using biomarkers of exposure such as metalothionein (Burger et al. 2000b) and food-web analysis using stable isotopes (Gaines et al. 2002).

5.5.1.4 Bats

Bats are the second most diverse order of mammals (after rodents) and constitute a substantial proportion of the mammalian biological diversity in the United States. Under current taxonomy there are 45 species of bats in the continental United States. Bats were not considered in the development of criteria for the Great Lakes Water Quality Initiative (GLWQI), which regarded upper-trophic-level piscivores as species most at risk. We believe that potential damage to bats should always be considered when assessing risk or deriving standards for waterborne contaminants, especially those that bioaccumulate. Given high throughput of potentially contaminated arthropod prey, combined with the relatively long life of bats, one might expect unusual bioaccumulation of stable contaminants relative to other small mammals. Their low reproductive rate (1 to 2 young per year) and long life span make them particularly vulnerable to bioaccumulative toxicants. Previously published total mercury levels in bats from the United States are analytically significant, and in bats roosting in abandoned mines, may be strikingly high. Data from northern California indicate significant differences between the *Myotis* species feeding on emerging aquatic insects from a mercury-polluted reservoir, *Antrozous* feeding on terrestrial insects near the reservoir, and *Plecotus* roosting in a mine nearby. The *Antrozous* data, although relatively low compared to the other 2, indicate a significant mercury exposure from a terrestrial source (Slotton et al. 1995).

Assays on bats in Japan in an area of mercury fungicide use revealed partitioning of Hg among various tissues with hair emerging as highest (Miura et al. 1978). Exposed cyanide-charged process water from heap leach gold mining operations has led to significant local bat mortality, demonstrating that bats will attempt to consume chemically contaminated water with potentially aversive odor and elevated pH (Clark et al. 1991; Clark and Hothem 1991). A bat of 10 g body weight, and 1 g/day food intake rate, if feeding on insects with total Hg concentrations such as those found in Clear Lake invertebrates, would be ingesting 5 to 20 times the mammalian Hg NOAEL used in the GLWQI model (Fenton 1992; USEPA 1993a; USEPA 1993b; USEPA 1995; USEPA 1997; Wolfe and Norman 1998).

5.5.1.5 Marine Mammals

Marine mammals encompass more than 120 living species within the orders Cetacea, Carnivora, and Sirenia, in addition to the sea otter (*Enhydra lutris*) and polar bear (*Ursus maritimus*) (Martin and Reeves 2002). Based on an assumption that anthropogenic changes to the global mercury cycle have had greater effects in coastal rather than open ocean waters, this discussion focuses on several species that are found more in coastal habitats. As with other contaminants in marine mammals, routes of mercury uptake of greatest concern are transplacental, via milk during suckling period (generally less significant), and via diet (Law 1996; Das et al. 2003); as with their terrestrial and freshwater counterparts, most dietary mercury is methylmercury. Once in the body, mercury can be transported to tissues that include the liver, kidney, muscle, skin, and hair. In general, most mercury in marine mammal muscle tissue is methylmercury, whereas liver and kidney tissue typically contain higher proportions of inorganic mercury (O'Hara et al. 2003).

Marine mammal mercury levels have been reported for 4 decades, and tissue analyses have involved the liver and, to a lesser extent, kidney, muscle, blubber, and hair (Law 1996; Wolfe et al. 1998; O'Shea 1999; Das et al. 2003; O'Shea and Tanabe 2003). The factors that can influence concentrations of mercury and other metals in marine mammals include species, age, sex, location, and predominant forage or prey, and concentrations will also depend on type and portion of tissue sampled, nutritive condition, and disease incidence (O'Hara et al. 2003). A number of studies have reported increasing mercury levels — and a decreasing percentage of MeHg in liver and kidney tissue — with age (Law 1996; Das et al. 2003), although this pattern has not been seen universally (see, for example, Atwell et al. 1998; Teigen et al. 1999). Species that would be potentially good indicators of changing mercury loadings to coastal environments include belugas, narwhals (*Monodon monoceros*), ringed seals (*Phoca hispida*), harbor seals, harbor porpoises (*Phocoena phocoena*), and polar bears (Law 1996; Wagemann et al. 1996; Wagemann et al. 1998). A number of methodological factors (including consideration of stranded animals vs. biopsies of free-ranging individuals) should also be taken into account in assessing the potential value as candidate biomonitor species for assessing responses to anthropogenic load changes.

5.5.2 BIRDS

5.5.2.1 Bald Eagle (*Haliaeetus leucocephalus*)

The bald eagle is distributed across North America. It is one of the most studied birds and its life history characteristics are well known. It is a tertiary predator and is indicative of food webs among all habitat types. The eagle's diet consists mainly of fish and other vertebrates associated with water bodies (Stalmaster 1987). Concentrations of mercury and other environmental contaminants have been measured since the 1960s across its range (Wiemeyer et al. 1984; Wiemeyer et al. 1993). The primary reason for using the eagle as a biosentinel species is its well-known life history, the ability to measure reproductive outcome accurately, the long-term database on both reproductive outcomes and concentrations of environmental contaminants, and its high visibility and appeal to humans. Concentrations of mercury have been reported in eggs (Wiemeyer et al. 1984; Wiemeyer et al. 1993), blood (Welch 1994; Evers et al. 2005), and feathers of both nestlings and adults (Wood 1993; Bowerman et al. 1994; Welch 1994; Wood et al. 1996; Bowerman et al. 2002). Archived feathers from museum collections have been used to determine exposure in the early 1900s (Evans 1993). Eagles have previously been identified as a useful biosentinel species for water quality and have been proposed as an indicator of Great Lakes water quality by the International Joint Commission (Bowerman et al. 2002).

5.5.2.2 Osprey (*Pandion haliaetus*)

The osprey is an obligate piscivore with a broad global distribution and a well-documented natural history (see many references in Poole 1989). The species benefits from an opportunistic foraging strategy and highly adaptable nesting habits. Ospreys have been regularly used as an indicator of contaminant exposure in regions such as the Great Lakes (Hughes et al. 1997), Chesapeake Bay (Rattner et al. 2004), Delaware Bay and surrounding regions (Clark et al. 2001), James Bay and Hudson Bay regions of Quebec (DesGranges et al. 1998), the Pacific Northwest (Elliott et al. 1998; Elliott et al. 2000), Oregon (Henny et al. 2003), and elsewhere. Mercury exposure has been reported for blood, adult and nestling feathers, and eggs (Wiemeyer et al. 1987; Anderson et al. 1997; Hughes et al. 1997; Cahill et al. 1998; Odsjo et al. 2004; Toschik et al. 2005). Mercury levels in blood of nestling ospreys have been found to be highly correlated with levels found in ingested prey, and are often less variable than other tissue types. Relationships between nestling osprey blood and feathers ($r^2 = 0.75$; DesGranges et al. 1998) are similar to those often reported in bald eagles (Wood 1993; Welch 1994; Weech et al. 2003). Adult feathers, often collected from the vicinity of nests and thought to reflect accumulation from the same area the previous year provide a significant excretory route for mercury and display higher mercury concentrations than nestling feathers (DesGranges et al. 1998). Osprey eggs are useful indicators of spatial and temporal trends in mercury exposure (Wiemeyer et al. 1984; Wiemeyer et al. 1987; Clark et al. 2001); however, eggs may not reflect contamination in the local food web in areas where they are laid before ice-out (DesGranges et al. 1999). Impacts of mercury on reproductive rates of ospreys have not been documented despite chronic exposure levels in some populations studied (i.e., impoundments in Quebec). Toxic effects may be greatest on post-fledge nestlings, because their feather molt no longer provides an excretory route for mercury.

5.5.2.3 Common Loon (*Gavia immer*)

This obligate piscivore is long-lived and during the breeding season is generally limited to a territory on a single lake (multiple lake territories are well known for lakes less than 60 acres; Piper et al. 1997; Piper et al. 2000). Other loon species, such as the yellow-billed loon (*Gavia adamsii*) and Pacific loon (*Gavia pacifica*), also well-represent mercury exposure on breeding territory lakes, whereas the widely roaming feeding habits of red-throated loons (*Gavia stellata*) make that species less useful for lake-specific exposure determinations. Common loon body mass varies dramatically by sex (average of 25% difference) and geographic area (contrasts of up to 50%) and therefore impact size of prey fish taken (Evers 2004). Generally, prey fish range from 10 to 25 cm and forage preferences on breeding lakes are yellow perch (*Perca flavescens*) (Barr 1986), centrarchids, and other species with a zigzag escape mechanism. Considerable efforts have been made to establish exposure profiles across North America (Evers et al. 1998; Evers et al. 2003b); and certain geographic high risk areas, such as Wisconsin (Meyer et al. 1998; Fevold et al. 2003), New England and New York (Evers et al. 1998; Evers et al. 2003a), eastern Ontario (Scheuhammer et al. 1998a), southern Quebec (Champoux 1996; Champoux et al. 2005), and the Canadian Maritimes (Burgess et al. 1998; Burgess et al. 2005). Because most lake systems are not connected to waterborne point sources, much of the mercury contamination represents atmospheric deposition. The use of blood and eggs has been shown to strongly reflect dietary uptake of fish Hg levels from breeding lakes (Meyer et al. 1995; Scheuhammer et al. 1998a; Evers et al. 2005). Large standardized databases (>3000 blood and >800 egg mercury levels (Evers and Clair 2005)), the ability to easily monitor marked individuals and recapture known individuals, high between-year breeding territory fidelity (Evers 2004), and new husbandry techniques (Kenow et al. 2003) make this species an important indicator for lakes and reservoirs.

5.5.2.4 Common Merganser (*Mergus merganser*)

This cavity-nesting duck is an obligate piscivore. It is well distributed across much of the northern United States and Canada. Breeding habitat includes both rivers and lakes. Only females incubate the generally 8 to 11 eggs laid. Dump-nesting, multiple females laying eggs in the same nest, is common and can result in greater than 20 eggs in a single nest. Well-established husbandry practices for waterfowl provide considerable potential for high-resolution laboratory studies for the common merganser. These characteristics, tied with the ability to direct nesting locations with the use of nest-boxes and the merganser's tendency to forage on fish and other small aquatic prey within a relatively small territory, make it a valuable indicator species for lakes, reservoirs, and large rivers (Timken and Anderson 1969; Mallory 1994; Champoux 1996; Ross et al. 2002; Champoux et al. 2005; Evers et al. 2005).

5.5.2.5 Seabirds

Seabirds have a wide distribution in marine, coastal, and inland aquatic environments, and many individual species have wide, worldwide, geographical ranges (common tern, black tern (*Chlidonia niger*), sooty tern (*Sterna fuscata*), herring gull, cormorant (*Phalacrocorax* sp.)). Seabirds are useful as bioindicators of coastal and marine pollution (Hays and Risebrough 1972; Gochfeld 1980; Walsh 1990; Furness and Camphuysen 1997). Seabirds, defined as birds that spend a significant proportion of their life in coastal or marine environments, are exposed to a wide range of chemicals because most occupy higher trophic levels, thus making them susceptible to bioaccumulation of pollutants. Selection of a particular species should depend on its life history strategy, breeding cycle, behavior and physiology, diet, and habitat uses (Burger et al. 2001). The relative proportion of time marine birds spend near shore, compared to pelagic environments, influences their exposure.

Multiple seabird species have been used as bioindicators for mercury, other metals, pesticides, chlorinated hydrocarbons, and petroleum products, particularly polyaromatic hydrocarbons (Burger and Gochfeld 2002). Because many species of seabirds eat mainly fish, indicators can be selected that are abundant locally and are at the top of their food chains. Eggs and feathers can be collected easily for most seabirds, and internal tissues can be collected where necessary. Some seabird species, such as the Leach's storm-petrel (*Oceanodroma leucorhoa*) have pelagic surface-feeding habits yet breed on offshore islands, thus serving as a potential bioindicator of trends in long-range atmospheric transport of mercury (Burgess and Braune 2002).

5.5.2.5.1 *Common Terns (Sterna hirundo)*

Common terns are widely distributed throughout the Northern Hemisphere, and into the Southern Hemisphere. They breed in a range of habitats from freshwater lakes to estuarine, coastal, and marine islands (Nisbet et al. 2002). They are long-lived seabirds (up to 30 years) that show a general fidelity to the same nesting area, and eat exclusively fish. They can be indicative of mercury exposure to predatory fish, and for other fish-eating birds. Data on status and trends in mercury and other contaminants exist for common terns from Europe (Becker and Sommer 1998), the Great Lakes (Stendell et al. 1976), and eastern North America (Burger and Gochfeld 1988; Burger and Gochfeld 2003). Thus, common terns are useful both on a temporal and spatial scale. Levels of mercury can be compared in parents and their eggs (Burger et al. 1999) in individually marked birds at different times, and in terns of different ages (Burger 1994). Data exist for mercury and other metals, PCBs (Hart et al. 2003), and DDT (Fox 1976).

Common tern tissues previously used for examining status and trends in mercury include feathers, eggs, and internal tissues. Common terns are migratory in most areas, requiring that information on time of arrival on the breeding grounds. Birds normally arrive 4 to 6 weeks before breeding (Burger and Gochfeld 1991) so levels in eggs represent local exposure. The feathers of young birds can be used as indicators of local exposure because parents provision chicks with fish from within a few kilometers.

5.5.2.5.2 Herring Gull (*Larus argentatus*)

Herring gulls are widely distributed in the Northern Hemisphere. They breed in a wide range of habitats, from freshwater lakes to estuarine and coastal environments, from sandy beaches to salt marshes, rocky ledges, cliffs, and trees (Pierotti and Good 1994; Burger and Gochfeld 1996a). They are long-lived seabirds (up to 40 years) that return to the same nesting colonies for many years. They eat a wide range of foods, from offal and garbage to carrion, invertebrates, and fish. They are useful as indicators because they are long-lived, breed on the same islands for many years, are very abundant and a pest species in many regions (making collection very easy), are amenable to laboratory experiments, some are nonmigratory, and there is an extensive literature on mercury levels. They have been used to assess status and trends for mercury and other metals in the Great Lakes and eastern North America (Burger and Gochfeld 1995; Gochfeld 1997) and in Europe. Herring gull eggs have been used to assess chlorinated hydrocarbons, particularly in the Great Lakes (Mineau et al. 1984; Oxynos et al. 1993; Pekarik and Weseloh 1998). Young herring gulls have also been used in the laboratory to examine neurobehavioral deficits (Burger et al. 2002) and to correlate tissue levels with effects for lead (Burger 1990; Burger and Gochfeld 2000a), making them a useful model for metal effects.

Herring gull tissues used include eggs, feathers and internal tissues. Herring gulls are migratory in most places, and nonmigratory in others, requiring assessors to understand the local ecology. Eggs and feathers of young birds are normally indicative of local exposure because parents arrive a month or 2 before egg-laying, and obtain all food for their chicks locally. Because parents have high nest site fidelity, the same individuals could be followed for several years.

5.5.2.5.3 Double-Crested Cormorant (*Phalacrocorax auritus*)

This species is the most widely distributed and abundant cormorant, found both on inland lakes and along all the coasts of North America. Their natural history is well-characterized and they are exclusively piscivorous, all features that promote their use as biosentinel species. Mercury residues and effects have been documented in cormorants in a number of studies (Henny et al. 1989, 2002; Burger and Gochfeld 1996b; Mason et al. 1997; Cahill et al. 1998; Sepulveda et al. 1998; Wolfe and Norman 1998a, 1998b; Burger and Gochfeld 2001). Cormorants are abundant and not a threatened species anywhere in their range, thereby simplifying sampling.

5.5.2.5.4 Belted Kingfisher (*Ceryle alycon*)

This short-lived species (3 years on average) is ubiquitous across much of North America and feeds exclusively on aquatic organisms. Prey size for adults varies from 5 to 12 cm. Kingfisher breeding territories generally encompass a 1- to 2-km area along a river, lake, or ocean shoreline from their sandbank burrow, and they occur across all general aquatic habitat types (marine, estuarine, riverine, and lake). Kingfishers are used to characterize waterborne mercury point sources (Baron et al.

1997) and in some cases atmospheric deposition sources (Evers et al. 2005). The strength of the kingfisher as an indicator species is its widespread distribution, ubiquitous aquatic habitat use, large and consistent clutch size (7 eggs), and ease of capture. However, multiple aquatic habitat types within a kingfisher territory can diminish the kingfisher's utility as an indicator of a target water body.

5.5.2.5.5 Egrets and Herons

Great blue herons (*Ardea herodias*) are among the upper-trophic-level piscivores at risk from environmental contaminants that bioconcentrate in aquatic food chains. Three large heron colonies located on the shore of Clear Lake in 1993 (2 in 1994) were useful for measuring mercury uptake by nestlings as a function of distance from the mercury source (Wolfe and Norman 1998). Great blue herons are widely distributed and often nest near contaminated sites, so there is substantial fund of comparative data (Quinney and Smith 1978; Hoffman 1980; Elliott et al. 1989; Fleming et al. 1985; Block 1992; Butler et al. 1995). Heron colonies in the western coastal states have been useful for monitoring contaminant concentrations in lakes, rivers, and estuaries. Mercury levels in heron tissue have been measured at a number of sites in the United States and elsewhere, providing a broad basis for comparison (Faber et al. 1972; Van Der Molen et al. 1982; Blus et al. 1985; Elliott et al. 1989). Heron chicks are siblicidal; the first-hatched nestling often kills or ejects from the nest subsequent hatchlings. These "excess" chicks can be collected for residue analysis without concern for population impacts.

The closely related great egret (*Ardea albus*) has a similarly wide distribution and life history, and has been successfully employed in mercury monitoring in the Florida Everglades (Hothem et al. 1995; Bouton et al. 1999; Duvall and Barron 2000; Rumbold et al. 2001; Spalding et al. 2000a, 2000b). Mercury uptake by great egrets has also been reported in China (Burger and Gochfeld 1993) and San Francisco Bay (Hothem et al. 1995).

5.5.2.6 Insectivorous Birds

Although piscivorous species are at higher trophic levels than insectivorous species, there is increasing concern that insectivorous songbirds also are at risk. In some studies, blood mercury levels in insectivorous songbirds exceed those of associated piscivores (Evers et al. 2005). Urban estuaries, freshwater wetlands, bogs, and acidic-montane habitats are among the potentially high-risk areas with increased MeHg availability (Evers et al. 2004). Species identified as suitable indicators of these habitats generally are those that are strictly insectivorous, have relatively small territories, and/or have known impacts. Estuarine species of interest include rails, especially clapper rails (*Rallus longirostris*) (Schwarzbach et al. 2000), saltmarsh sharp-tailed sparrows (*Ammodramus caudacutus*), and seaside sparrows (*Ammodramus maritimus*). Montane bird communities in the Northeast appear to have elevated blood mercury levels (Rimmer et al. 2005), which is of particular concern for the relatively endemic Bicknell's thrush (*Cathartus bicknelli*). This finding also indicates that strictly terrestrial environments can contain available MeHg at levels that may put nonaquatic species as risk. Western montane habitats may be best monitored with the American dipper (*Dolichonyx oryzivorus*). Use of aquatic habitat generalists as indicators of MeHg availability is needed. Tree swallows (*Tachycineta bicolor*) quickly colonize new areas, use nest boxes that provide ease of accessibility, and are part of recent experimental dosing studies (Echols et al. 2004; Custer et al. 2005; Mayne et al. 2005), as well as having documented exposure and uptake of metals (Kraus 1989; Nichols et al. 1990; Barlow 1993; Bishop et al. 1995a). Although other swallow species are less responsive to artificial structures, experiments designed to measure environmental mercury levels can opportunistically employ these ubiquitous insectivores as well (Grue et al. 1984; King et al. 1994; Ellegren et al. 1997). European starlings (*Sturnus vulgaris*) also utilize nestboxes (Wolfe and Kendall 1998), and a protocol exists for the use of starling nestboxes in toxicity studies (Kendall et al. 1989). At Clear Lake, California, where mercury enters the lake from a point source (an abandoned mine), red-winged blackbird (*Agelaius phoeniceus*), Brewer's blackbird (*Euphagus cyanocephalus*), and cliff swallow (*Hirundo pyrrhonota*) nestlings were sampled to confirm that these insectivorous birds were exposed to Hg and MeHg and to see if an effect of distance from the mine was evident in this lower trophic level. Samples of insect food collected from passerine foraging areas contained 0.01 to 0.420 ppm total mercury. Total mercury residues in nestlings were 0.018 to 0.03 ppm in brain; 0.094 to 0.322 ppm in feathers, for all ages of all 3 species (Wolfe and Norman 1998). A similar multi-species approach was completed at a Superfund site on the Sudbury River (Massachusetts) and was used to characterize mercury exposure and risk to insectivorous birds (Evers et al. 2005). Same-site comparisons show blood mercury levels in song and swamp sparrows (*Melospiza melodia*) and (*M. georgiana*) consistently exhibited greater blood mercury levels than yellow warblers (*Dendoica petechia*) and common yellowthroats (*Geothlypis trichas*). Blood mercury levels were highest in red-winged blackbirds (>1.2 ppm, ww) and northern waterthrush (*Seiurus noveboracensis*) (>1.6 ppm, ww).

5.5.3 REPTILES AND AMPHIBIANS

5.5.3.1 Reptiles

5.5.3.1.1 Alligators

Alligators (*Alligator mississippiensis*) are top-level predators that live at the water/land interface in the southeastern part of the United States. They are useful bioindicators because they eat large fish, turtles, and even egrets, and can be indicative of exposure of other top-level carnivores, including wading birds, hawks, and humans. They have been used as bioindicators of organochlorines and endocrine disruptors (Heinz et al. 1991; Guillette et al. 1994; Guillette et al. 1996) and heavy metal contamination, including mercury (Delany et al. 1988; Heaton-Jones et al. 1997; Yanochko et al. 1997). There are studies of mercury in alligators from many places within the Southeast (Ruckel 1993; Yanochko et al. 1997; Brisbin et al. 1998), making them useful for this geographical region. As well as internal tissues, tail and skin have been used as bioindicators of exposure; skin gave the highest correlation with mercury levels in internal tissues (Burger et al. 2000a).

5.5.3.1.2   Water Snakes

Water snakes (*Nerodia* sp.) are commonly distributed throughout the United States east of the Rockies, and occur in rivers, lakes, streams, marshes, and adjacent uplands. They are carnivorous, foraging on a wide range of invertebrates, amphibians, and fish. They are useful because they are top-level predators, there is a vast literature on their ecology and behavior, and there is more information on contaminants in this snake than any other (Campbell and Campbell 2001). They thus provide information on the land/water interface. They have been used as bioindicators in many regions, including the Northeast (Burger et al. 2004), the Southeast (Campbell et al. 1998), and the Great Lakes Basin (Bishop and Rouse 2000), and in the closely related diamondback water snakes (*Nerodia rhombifer*) and blotched water snakes (*Nerodia erythrogaster*) in Texas (Clark et al. 2000).

5.5.3.1.3   Turtles

The snapping turtle (*Chelydra serpentina*) is distributed across North America east of the Rockies. It is well studied and many of its life-history characteristics are known. It is a tertiary predator and scavenger of aquatic systems and is indicative of food webs among freshwater habitat types. The snapping turtle's diet consists mainly of fish and other vertebrates associated with aquatic systems (Bishop et al. 1995b). Concentrations of mercury and other environmental contaminants have been collected from the 1990s to present across its range (Bishop et al. 1998; Golet and Haines 2001). The primary reason for the turtle as a biosentinel species is its well-known life history, the ability to collect eggs from the wild and hatch them in captivity, the long life span and small home ranges of turtles, and widespread distribution and relative intolerance to human activities. Concentrations of mercury have been reported in eggs and in tissues of both nestlings and adults (Meyers-Schoene and Walton 1990; Meyers-Schone et al. 1993; Bonin et al. 1995; Bishop et al. 1998; Golet and Haines 2001; Ashpole et al. 2004). Turtles were previously proposed as a useful biosentinel species for water quality and are continuing in the development stage (Nisbet 1998).

5.5.3.2   Amphibians

Very little data exist on mercury levels in amphibians. Recent efforts by Bank et al. (2005) indicate that the northern 2-lined salamander (*Eurycea bislineata bislineata*) is a top indicator of MeHg in stream ecosystems. Bullfrogs (*Rana catesbeiana*) are abundant and widely distributed. In the West, they are pests, making their use in toxicity studies particularly attractive in that part of the country. They have been used for mercury studies by several investigators (Birge and Just 1973; Tsuchiya and Okada 1982; Sillman and Weidner 1993; McCrary and Heagler 1997; Burger and Snodgrass 1998; Rowe et al. 1998).

Table 5.1 summarizes the species listed above and ranks them as potential bioindicators of mercury contamination according to the characteristics discussed, from 1 (lowest) to 3 (highest), based on the assessments above and the best professional judgement of the authors of this chapter.

TABLE 5.1
Desirable characteristics of candidate biomonitor species and ranking according to characteristic

	Sensitive, indicative of change	Broad distribution with accompanying data	Easily measured and readily observable	Well-known ecology and life history	Suitable for lab studies	Important to humans	Economical/cost effective	Well-developed and usable with existing data	Common enough not to impact populations
Mammals									
Mink	3	3	3	3	3	3	3	3	3
Raccoon	3	3	2	3	2	3	3	3	3
River otter	3	1	2	2	1	2	2	2	2
Bats	3	1	1	1	2	1	1	2	2
Beluga whale	2	3	2	3	1	3	1	3	3
Narwhal	2	2	2	3	1	2	1	3	3
Ringed seal	3	3	3	3	2	2	3	3	3
Harbor seal	3	3	3	3	2	2	3	2	3
Harbor porpoise	3	3	3	3	2	2	2	2	3
Polar bear	2	2	2	3	1	3	2	3	3
Birds									
Common loon	3	2	3	3	2	3	3	3	3
Double-crested cormorant	3	2	2	3	1	2	2	2	3
Seabirds	2	2	2	2	1	2	2	2	2
Great blue heron	2	1	3	3	1	2	2	2	3
Great egret	2	1	3	3	1	3	3	3	3
Common and hooded merganser	3	2	3	3	3	2	3	2	3
Bald eagle	3	2	1	3	1	3	3	3	1
Osprey	2	3	2	3	1	3	3	2	2
Rail spp.	3	3	1	1	2	1	1	1	2
Willet	2	1	2	1	1	1	2	1	1
Herring gull	3	2	3	3	3	1	3	3	3
Common tern	3	2	3	3	3	3	3	3	3
Belted kingfisher	3	2	3	3	3	2	2	2	3
Insectivorous songbirds									
Tree swallow	3	3	3	3	3	3	3	2	2
Bicknell's and wood thrush	3	2	2	3	1	1	1	2	3
European starling	2	3	3	3	3	1	3	2	3
Louisiana and northern waterthrush	3	1	3	2	1	1	2	1	3
Red-winged blackbird	2	3	3	3	2	3	3	2	3
Sharp-tailed and seaside sparrow	3	1	3	2	1	2	3	2	2

TABLE 5.1 (continued)
Desirable characteristics of candidate biomonitor species and ranking according to characteristic

	Sensitive, indicative of change	Broad distribution with accompanying data	Easily measured and readily observable	Well-known ecology and life history	Suitable for lab studies	Important to humans	Economical/cost effective	Well-developed and usable with existing data	Common enough not to impact populations
Reptiles									
Crocodile	1	1	3	1	3	1	1	1	1
Alligator	1	1	1	3	3	2	2	3	2
Snapping turtle (East)	2	2	1	3	3	2	2	2	3
Red-eared slider (West)	2	3	2	3	3	1	2	2	3
Water snake	3	2	2	3	3	3	3	2	3
Lizards									
Sceloporus	2	1	1	2	3	1	1	1	2
Amphibians									
Bullfrog	2	2	2	3	3	1	3	2	3
Two-lined salamander (East)	3	1	2	1	1	1	2	1	1
Slender salamander (West)	3	1	2	1	1	1	2	1	1

Based on the scoring in Table 5.1 (summing scores for each species), candidate bioindicator species can be ranked within a taxonomic group according to suitability for a mercury monitoring program for North America:

- Terrestrial and aquatic mammals, from highest to lowest: mink, raccoon, river otter, bats.
- Marine mammals: ringed seal, harbor seal, harbor porpoise, beluga whale, narwhal, polar bear.
- Birds: common loon, common tern, common merganser, herring gull, tree swallow, red-winged blackbird, European starling, belted kingfisher, great egret, great blue heron, bald eagle, double-crested cormorant, other seabirds.
- Reptiles: water snake, alligator, snapping turtle, red-eared slider, *Sceloporus* sp.
- Amphibians: bullfrog, 2-lined salamander, slender salamander.

5.5.4 OTHER POTENTIAL INDICATORS

5.5.4.1 Albatrosses

Although not used extensively, albatrosses should prove useful because they are very long-lived (60 years or more), have high fidelity to their nest sites, and bioaccumulate contaminants over time. Disadvantages include a wide feeding range (often several hundred kilometers from the breeding colony), and often variability in feeding ranges. Nonetheless, they are vulnerable to contamination, and were used as an indicator of lead poisoning on Midway Island. Baseline data on mercury and other metals exist from Midway and elsewhere that could be used for future assessment (Thompson et al. 1993; Kim et al. 1996; Hindell et al. 1999; Burger and Gochfeld 2000b).

5.5.4.2 Hawks

The concentration of mercury in terrestrial ecosystems has been determined through the use of tissue samples from hawks and other birds of prey. Feathers of the northern goshawk (*Accipiter gentilis*) have been used to determine the concentration of mercury in Sweden (Wallin 1984). Feathers of other hawks, including the red-tailed hawk (*Buteo jamaicensis*), sparrowhawks (*Accipiter nisus*), eagle owls (*Bubo bubo*), gyrfalcons (*Falco rusticolus*), and merlin (*Falco columbarius*), have also been used. Hawks occupy the tertiary predator role of terrestrial food webs; and as with other semi-aquatic predatory birds, they are useful indicators of bioaccumulative compounds in the environment. With the lack of any well-developed indicator species for the terrestrial system, monitoring projects using hawks and other birds of prey should be developed.

5.5.5 IDENTIFICATION OF INDICATORS THROUGH DEVELOPMENT OF WATER QUALITY CRITERIA FOR WILDLIFE

Development of water quality criteria (WQC) in the United States is an additional process that has involved identifying wildlife indicators for mercury (and other contaminant) exposure. As part of the development of uniform water quality standards for the Great Lakes states, the USEPA derived water quality criteria for protection of wildlife for 4 pollutants, including mercury (U.S. Code of Federal Regulations, 40 CFR Part 132). The approach involved both an exposure and a hazard component, and derivation of criteria values for 5 species of concern (eagles, herring gull, kingfisher, mink, and otter). The criteria were converted to total mercury concentrations in water, and the geometric mean for avian species yielded a value of 1.3 ng/L as the wildlife value (reviewed in Nichols et al. (1999)). A similar approach in the USEPA Mercury Study Report to Congress for 5 species (with kingfisher replacing osprey in the group above) yielded wildlife values ranging from 0.6 ng/L for kingfisher to 1.8 ng/L for eagles (USEPA 1997). (There is currently no formal national WQC guideline in the United States explicitly developed for protection of wildlife from mercury, or any other chemical.)

There have also been efforts in individual states in the United States to develop WQC for wildlife. For example, Maine is currently developing a mercury wildlife value, and research on loons in the region has been used in support of that effort. Based on adult loon blood levels leading to impairments in fledged young, and default bioaccumulation factors, a wildlife value (expressed on a total mercury concentration basis) was derived (Evers et al. 2003a). In an additional assessment by the U.S. Fish and Wildlife Service of the protectiveness of the USEPA's MeHg criterion for protection of human health (USEPA 2001) to also protect wildlife, in California it was found that under the highest trophic level approach, the criterion (0.3 μg/g in fish) would not offer protection for 2 federally listed species (California least tern and Yuma clapper rail) (Russell 2003).

Despite these efforts, there are recognized difficulties in deriving a single water quality criterion for mercury, given the number of factors impacting MeHg formation and bioaccumulation. These issues have been raised previously in the literature (Kelly et al. 1995; Meyer 1998), and are reviewed again in Chapters 3 and 4. Moore et al. (2003) addressed limitations in this approach by developing a water quality criteria model that incorporated factors impacting bioavailability, methylation rates, and bioaccumulation in aquatic systems, based on an analysis of data from 41 lakes. Based on the use of mink mortality as the endpoint and on a probabilistic model relating MeHg levels in water to fish levels, a model allowing for site-specific inputs was developed. This model will need to be evaluated with larger data sets across a wide variety of watersheds and water-body biogeochemical characteristics to determine its broader applicability.

5.6 TISSUE AND OTHER SAMPLES

5.6.1 HAIR

Hair has been recognized as a bioaccumulator of heavy metals since before the turn of the century. It was used in forensic studies before environmental studies because of the earlier interest in forensic matters. As early as 1908, there were reports of arsenic in horsehair near a smelter in Montana. The use of hair for determining body burden of mercury and other metals has been recognized for many years (Aoki 1970; Eyl 1971; Albanus et al. 1972; Birke et al. 1972; Roberts et al. 1974). The development of simpler and more accurate detection methods in the 1960s and 1970s, coupled with interest in environmental monitoring, led to its widespread use as a bioindicator. Jenkins (1980), in conducting an USEPA review of biological monitoring, concluded that hair is a good bioindicator for certain elements. Huckabee et al. (1973), working with coyotes and rodents (e.g., mice, voles, chipmunk, porcupine), first suggested a strong positive correlation between environmental mercury and hair mercury. They estimated that wildlife hair levels exceeding a mean of about 0.6 ppm may be evidence of an abnormally high occurrence of mercury in the environment, and concluded that hair may serve as an effective monitor of environmental mercury. Since then, data on hair mercury in wild populations have been reported for bobcats, raccoons, opossum, fox, deer, squirrels, mink, otter, bear, wild boar, mountain goat, elk, muskrat, beaver, and panther. Hair mercury levels are often higher than levels in other tissues and have been significantly correlated ($p < 0.01$) with mercury concentrations in other tissues for a number of species (Cumbie 1975b). Another important advantage of the use of hair as a bioindicator is the existence of inter-laboratory validation programs for hair analysis such as that initiated and maintained by Environment Canada, an outgrowth of the use of hair in human forensic studies. Most investigators report that hair mercury strongly correlates with concentrations in internal tissues (Farris et al. 1993; Nielsen et al. 1994; Evans et al. 2000; Mierle et al. 2000). However, a recent study on mink in South Carolina and Louisiana found that no correlation existed between Hg concentrations in hair and other tissues, or among Hg concentrations from hair taken from 3 locations on the same carcass (Tansy 2002). Additional standardization of hair collections from carcasses and assessment of relationships between hair and tissue Hg concentrations is necessary to ensure comparability among geographic locations.

5.6.2 FEATHERS

Feathers are useful indicators of heavy metal contamination, particularly for mercury. However, they are not useful for other contaminants. Birds sequester heavy metals in their feathers during feather formation, and the intact feather is a record of exposure during that time. The proportion of the body burden that is in feathers is relatively constant for each metal, and a relatively high proportion of the body burden of certain metals is stored in the feathers (Burger 1993). For mercury, about 70% (Honda et al. 1986) to 93% (Braune and Gaskin 1987) of the body burden is in feathers, and greater than 95% of the mercury in feathers is MeHg (Thompson and Furness 1989a, 1989b). Monteiro et al. (1998) demonstrated that there is a high correlation between levels of mercury in the diet of seabirds and levels of mercury in their feathers; thus, feathers can be used as indicators of food chain effects. Evers et al. (1998) showed that inter-compartmental relationships, such as between loon blood and feathers, are complex and are heavily influenced by lifetime body burdens and throughput; in the continental loon study, the annual increase for males was 9%.

Feather collection is a nondestructive, noninvasive method of obtaining mercury exposure. It is thus especially useful for threatened or endangered species (Gochfeld and Burger 1998). Feathers can also be used to examine age and gender effects (Thompson et al. 1991). Because feathers are stable, and do not break down over time, they can be archived for later analysis, providing the opportunity to analyze temporal trends using the same instrumentation. Feathers in museum collections have proven particularly useful for examining changes in mercury levels over centuries (Berg et al. 1966; Walsh 1990; Thompson et al. 1992). Furthermore, there are several archived collections of feathers for several species in university collections, as well as in museums. Care must be taken to consider the preservation method for feathers, as mercury and arsenic were sometimes used.

The use of feathers as an indicator of mercury exposure requires understanding the life cycle and ecology of each species. Because some species migrate, the location of exposure can be problematic. However, this disadvantage can be eliminated by using nonmigratory species and young birds nearly ready to fledge (Burger 1993). Local exposure can be compared to exposure on the wintering grounds in adults that incubate for at least 3 weeks by collecting breast feathers at the beginning of incubation, and then collecting the regrown feathers after 3 weeks (Burger et al. 1992).

5.6.3 EGGS

Eggs are important indicators of adult exposure and effects of environmental contaminants on embryos, usually the most sensitive stage for effects (Rudneva-Titova 1998; Bellas et al. 2001; Kiparissis et al. 2003). Eggs of reptiles and amphibians (Bonin et al. 1995; Burger et al. 2000a), fish (McMurtry et al. 1989; Rudneva-Titova 1998; Tatara et al. 2002), invertebrates (Canli and Furness 1993), and birds (Scheuhammer et al. 2001; Champoux et al. 2002; Evers et al. 2003a) have been collected from the field to determine exposure to environmental contaminants. Eggs have been collected from the field and incubated in laboratories to determine the effects of contaminants on reproduction, survival, teratology, biochemical markers, and eggshell quality (Burger and Gochfeld 1988; Leonzio and Massi 1989; Eriksson et al. 1992; Bishop et al. 1998; de Solla et al. 2002; Cifuentes et al. 2003; Evers et al. 2003b). Eggs are collected either as fresh eggs (Becker et al. 1993) or as abandoned or addled eggs (Koivusaari et al. 1980; Steidl et al. 1991; Ruelle 1992) for residue studies. Maternal transfer of mercury to eggs represents body burdens and dietary uptake. In loons, eggs and female blood mercury levels were strongly correlated ($r^2 = 0.79$), establishing recent dietary uptake as the pathway most responsible (Evers et al. 2003b).

5.6.4 ORGANS

5.6.4.1 Blood

The blood compartment provides a method for determining recent MeHg uptake. Nearly all mercury (>95%) in the blood of mammals and birds is in the form of MeHg (Thompson 1996). The half-life of MeHg in the blood of juvenile common loons is 114 days (Fournier et al. 2002). However, if an organism travels from a low mercury risk site to a high mercury risk site, mercury levels from the latter site will be reflected. For example, adult common loons migrating from marine areas (blood mercury levels range from 0.1 to 1.1 ppm, ww) to interior freshwater systems (blood mercury levels range from 0.3 to 9.5 ppm, ww) are moving from generally low to high risk areas and therefore their blood mercury levels quickly reflect the increase in MeHg availability. Further evidence is the strong relationship between prey fish and adult blood mercury levels on breeding lakes ($r^2 = 0.80$) (Evers et al. 2004).

5.6.4.2 Brain

Brain is a key tissue to analyze for mercury concentration because it is the site of MeHg toxicity. The neurotoxic effects of MeHg in adult mammals include ataxia, difficulty in locomotion; neurasthenia, a generalized weakness; impairment of hearing and vision; tremor; and finally loss of consciousness and death (Eaton et al. 1980; Wren et al. 1987b; Heinz 1996). Methylmercury damages primarily the cerebellum and cerebrum (Chang 1990). Methylmercury accumulates preferentially in the posterior cortex. Lesions in the cerebral and cerebellar cortex accompany these clinical signs. Necrosis, lysis, and phagocytosis of neurons result in progressive destruction of cortical structures and cerebral edema. O'Connor and Nielsen (1981) found necrosis, astrogliosis, and demyelination in the cerebral and cerebellar cortex of the otters that received 0.09, 0.17, and 0.37 mg/kg/d MeHg for 45 to 229 days. In adult mammals, MeHg is preferentially taken up by glial cells; these appear particularly susceptible to MeHg damage (Takeuchi 1977). Low concentrations (10^{-5} M) of MeHg inhibit the ability of cultured rat brain astrocytes to maintain a transmembrane K^+ gradient, resulting in cellular swelling (Aschner et al. 1990). Methylmercury is readily transferred across the placenta and concentrates selectively in the fetal brain. Hg concentrations in the fetal brain were twice as high as in the maternal brain for rodents fed MeHg (Yang et al. 1972).

In birds, a brain mercury concentration of less than 2 ppm wet weight was associated with reduced egg laying, and impaired nest and territory fidelity in common loons (Barr 1986). Black duck embryos with brain mercury concentrations of 4 to 6 ppm failed to hatch (Finley and Stendall 1978). Brain mercury concentrations of 20 ppm caused 25% mortality in mercury-exposed zebra finches (Scheuhammer 1988).

5.6.4.3 Liver

Liver is 1 of the tissues most frequently analyzed for contaminant residue in wildlife, but maybe 1 of the least useful because of the poor correlation between liver mercury concentration and effects, and because of the tendency of the liver to accumulate mercury over time (Stewart et al. 1999; Scheuhammer et al. 2001). Liver is a major site of demethylation; therefore, the proportion of liver mercury present as MeHg is not representative of exposure to MeHg. Moreover, most mercury in liver is bound to metallothionein or other sulfydryl-bearing proteins, which immobilize it (Medinsky and Klaassen 1996; Yasutake et al. 1997; Aschner 1999). Therefore, liver mercury residue values must be used with caution, and only when more suitable tissues are unavailable.

5.6.4.4 Muscle

Most of the mercury in muscle is present as MeHg. Barr (1986) reported that adult loons from a site closest to a mercury source with 1.87 ppm Hg in fish, resulted in a muscle Hg concentration of 4.57 ppm ww and a brain Hg concentration of 1.49 ppm. Those loons had only 20% of territories successful (Barr 1986). Mallard hens receiving 0.5 ppm dietary mercury for 18 months had a muscle Hg concentration of 0.82 ppm compared with brain Hg = 0.50 ppm. Mallard hens receiving 3 ppm for the same period had 5.01 ppm in muscle and 4.57 in brain (Heinz 1976). Based on these assessments, muscle Hg concentration is more representative of brain Hg concentration and correlates better with effect than the more commonly measured liver residue. It is also possible to sample muscle tissue nonlethally via biopsy, in situations where internal body tissue is needed in addition to or instead of feather or fur (Anderson 1997; Dickinson et al. 2002).

5.6.4.5 Kidney

Kidney residue analysis is used primarily for inorganic mercury because that is where inorganic mercury exerts its toxic action (Nicholson and Osborn 1983, 1984; Goering et al. 1992). Analysis of mercury in kidney tissue was, along with liver, relatively common in earlier field studies of mercury contamination in bird populations (e.g., reviewed in Heinz 1996; Thompson 1996). The kidney is a major repository of inorganic mercury in both birds and mammals, and a site where toxicity can be manifested (see review in Wolfe et al. 1998). Both liver and kidney mercury analyses have been common in studies of marine mammals (e.g., Law 1996; O'Shea 1999). As with liver data, interpretation of kidney mercury levels in marine mammals can be challenging, due to the potential for both demethylation and sequestration (e.g., as a mercury selenium compound) in the organ (e.g., Das et al. 2003; O'Hara et al. 2003).

5.7 PHYSIOLOGICAL, CELLULAR, AND MOLECULAR BIOMARKERS

Subcategories of bioindicators at suborganismal levels are generally referred to as biomarkers. Biomarkers may reflect either exposure, effect, or susceptibility; of these, biomarkers of effect are both more valuable and more difficult to find. An *ideal* biomarker of mercury effect would be:

- Predictive (that is, it would reflect very early response to contaminant impact, before functional damage has occurred) (Kreps et al. 1997)
- Specific for Hg or MeHg
- Based on samples that can be obtained nonlethally and noninvasively, and that can be sampled repeatedly in the same individual
- Validated in wildlife species

Reported impacts of mercury on individuals can be categorized into physiological/functional, morphological, behavioral, reproductive, and demographic.

Recent advances in molecular ecology and genetic analysis have increased our ability to examine the effects of mercury. The replicability of genetic assays makes them particularly attractive.

In this section, we discuss reported endpoints of mercury exposure and effect at the sub-organismal level, and evaluate them as to their potential usefulness as biomarkers, either alone or in combination. Although numerous mercury effects have been documented in the ongoing effort to explicate the biochemical mechanism of mercury toxicity, no known mercury endpoint fulfills the desirable criteria for a biomarker. Given current knowledge, the most promising approach may be to identify not a single endpoint, but to combine a suite or panel of easily measured nonspecific endpoints that occur together, and may in aggregate indicate MeHg toxicity (Basu et al. 2006). A meta-analysis of existing published data may be a fruitful first endeavor toward discovering the best candidates for such a fingerprint (Bailar 1995; Ioannidis and Lau 1999). Next, because most sub-organismal endpoints are studied *in vitro*, suitable mechanistic methods must be employed to predict which are the most likely candidates to be verified *in vivo* (Ponce et al. 1998; Lewandowski et al. 2001).

The work of Hoffman and Heinz illustrate how this approach can be applied in wildlife species. They exposed mallards to MeHg, with or without selenium co-exposure, and then measured hematocrit, hemoglobin, and plasma chemistries; reduced and oxidized glutathione and activities of several glutathione pathway enzymes, G-6-PDH activity, brain lipid peroxidation and thiobarbituric reactive substances as an indicator of oxidative stress. These laboratory measurements constituted a profile of cellular and biochemical MeHg and MeHg/Se effects, which they then compared to those from wild-caught waterfowl from San Francisco and Suisun Bays, and to tissue residues of mercury and selenium (Heinz and Hoffman 1998; Hoffman and Heinz 1998; Hoffman et al. 1998).

More recently, Henny et al. (2002) have applied this same profile approach to double-crested cormorants, snowy egrets, and black-crowned night herons from the Carson River system, demonstrating its across-species applicability. Henny and co-workers expanded the profile to include histopathological parameters.

Specific biomarkers of mercury toxicity may well emerge from investigations that employ technologies from outside the toxicological sciences to elucidate some of the unique characteristics of mercury's toxic mechanisms. For example, mercury is 1 of the few environmental contaminants known to affect the visual system (Fox and Boyes 2001). Electroretinography was used to measure visual changes in great egrets and juvenile double-crested cormorants that had been experimentally dosed with MeHg (Loerzel et al. 1996; Loerzel et al. 1999). Although the concentrations of mercury in the ocular tissue were very low, retinal morphology was affected by systemic exposure to MeHg.

Wildlife toxicologists should be attuned to developments in human health mercury, as assays that have been used successfully on humans may be suitable or adaptable for other vertebrate species. Echeverria and co-workers (Echeverria et al. 2005, 2006; Heyer et al. 2006) have characterized a gene encoding coproporphyrinogen oxidase, a gene in the heme biosynthetic pathway. Polymorphism in this gene predicts differential response to elemental mercury exposure in human subjects. Plans to modify this assay for other mercury species in matrices from wildlife are under way.

Table 5.2 summarizes sub-organismal endpoints reported in the literature, roughly categorized as hematological, enzymatic, immunological, and neurological, although necessarily, there is much overlap.

5.7.1 WHAT IS IN THE PIPELINE? FUTURE AND PROMISING BIOMARKERS

Current work directed at elucidating the mechanism of mercury toxicity will be a fertile area of research into potential sub-organismal biomarkers. Techniques including

TABLE 5.2
Mercury endpoints that may be useful as biomarkers

Endpoint	Tissue or cell	Species	Nonlethal?	Effect	Ref.
Enzyme activity					
Glutathione (GSH)	Liver	Mice	No	Decreased	Balthrop and Braddon (1985)
	Liver	Greater scaup, surf scoter, ruddy duck	No	Decreased	Hoffman et al. (1998)
GSH-S-transferase	Liver	Cormorant	No	Decreased	Henny et al. (1998)
	Liver	Snowy egret	No	Increased	Henny et al. (1998)
	Liver	Great egret	No	Increased	Hoffman et al. (2005)
GSH peroxidase	Liver	Surf scoter, ruddy duck	No	Decreased	Hoffman et al. (1998)
	Liver, plasma	Mallard duck	No	Decreased	Hoffman and Heinz (1998)
	Liver	Great egret	No	Decreased	Hoffman et al. (2005)
GSSG/GSH	Liver	Mallard duck	No	Increased	Hoffman and Heinz (1998)
Na$^+$/K$^+$-ATPase	Ethyrocytes	Rat	N/A	In vitro, not in vivo	Maier and Costa (1990)
Lactic dehydrogenase	Serum	Harp seal	Yes	Increased	Ronald et al. (1977)
Alkaline phosphatase	Serum	Harp seal	Yes	Increased	Ronald et al. (1977)
gamma-GC	Kidney	Mouse	No	Increased	Yasutake and Hirayama (1994)
glucose-6-phosphate dehydrogenase (G-6-PDH)	Liver	Cormorant	No	Decreased	Henny et al. (1998)
	Liver	Snow egret	No	Increased	Henny et al. (1998)
	Liver	Surf scoter	No	Decreased	Hoffman et al. (1998)
	Liver	Mallard duck	No	Decreased	Hoffman and Heinz (1998)
gamma-Glutamylcysteine synthetase	Brain	Mice	No	Upregulated in resistant regions	Li et al. (1996)

TABLE 5.2 (continued)
Mercury endpoints that may be useful as biomarkers

Endpoint	Tissue or cell	Species	Nonlethal?	Effect	Ref.
Acetylcholinesterase	Plasma	Coturnix quail	Yes	Decreased	Dieter and Ludke (1975)
Benzopyrene monooxygenase	Midgut gland, haemolymph, gills	Carcinus crab		Induced	Fossi et al. (1996)
Blood					
Hemoglobin	Whole blood	Mallard	Yes	Decreased	Hoffman and Heinz (1998)
	Whole blood	Mouse	Yes	Decreased	Shaw et al. (1991)
Erythrocytes	Whole blood	Harp seal	Yes	Decreased	Ronald et al. (1977)
White blood cell count	Whole blood	Harp seal	Yes	Increased	Ronald et al. (1977)
Inorganic phosphorus	Whole blood		Yes	Decreased	Elbert and Anderson (1998)
Monoamine oxidase activity	Platelets	Rats	Yes	Decreased	Chakrabarti et al. (1998)
Apoptosis, via reactive oxygen species (ROS)	Peripheral monocytes	Human	Yes	Increased	InSug et al. (1997)
Blood cell ratio	Whole blood	Rainbow trout	Yes	No change	Niimi and Lowe-Jinde (1984)
		Great blue heron, Cliff swallow, Red-winged blackbird	Yes	See text	Wolfe and Norman (1998)
Packed cell volume	Whole blood	Rainbow trout	Yes	Increased Decreased	Wobeser (1975) Rogers and Beamish (1981)
		Bass	Yes	Decreased	Dawson (1982)
		Flounder	Yes	Decreased	Calabrese et al. (1975)
		Mallard	Yes	Decreased	Hoffman and Heinz (1998)
		Mouse	Yes	Decreased	Shaw et al. (1991)

TABLE 5.2 (continued)
Mercury endpoints that may be useful as biomarkers

Endpoint	Tissue or cell	Species	Nonlethal?	Effect	Ref.
Packed cell volume	Whole blood	Great egret	Yes	Decreased	Spalding et al. (2000b)
Serum cholesterol	Serum	Quail	Yes	Decreased	Leonzio and Monaci (1996)
Cortisol	Plasma	Rainbow trout	Yes	Increased	Bleau et al. (1996)
	Plasma	Walleye	Yes	Suppressed	Friedmann et al. (1996)
Thyroxine (T4) and	Plasma	—			Bleau et al. (1996)
Triiodothyroxine (T3)	—	—			
Immune function					
Natural killer cell	Lymphocytes	Rat (prenatal and lactation exposure)	Yes	Reduced 42%	Ilback et al. (1991)
Oligoclonal CD4+ T-cell response	Spleen, thymus	Balb/c mouse	No	Increased in splenic, not in thymic	Heo et al. (1997)
White blood cell Phagocytosis	Whole blood	Chicken	Yes	No change	Holloway et al. (2003)
	Isolated cells			Depressed	
	Whole blood	Loon		No change	
Calcium influx	Immune	Teleost fish	—	Increased	Burnett (1997)
Monocyte phagocytosis	Monocytes, *in vitro*	Human	*	Decreased	InSug et al. (1997)
Apoptosis, via ROS				Increased	InSug et al. (1997)
Macrophage centers	Macrophages of liver, kidney and spleen	Pike (*Esox lucius*)	No	Increased	Meinelt et al. (1997)
Hemagglutination titers	Whole blood	Mouse	Yes	Suppressed	Blakeley et al. (1980)
B-cell production	Whole blood	Mouse	Yes	Suppressed	
t-cell apoptosis	t-cells, *in vitro*	Human	*	Increased	Shenker et al. (1998)
Tyrosine phosphorylation	Lymphocytes	Rat	No	Induced	Allen et al. (2001)

TABLE 5.2 (continued)
Mercury endpoints that may be useful as biomarkers

Endpoint	Tissue or cell	Species	Nonlethal?	Effect	Ref.
Genetic					
Induction of Gadd45 and Gadd153 genes	Embryonic neuronal cells	Mouse	No	Induced expression of DNA damage genes	Ou et al. (1999)
Gene expression of monocyte chemotactic protein	Peripheral blood mononuclear cells	Human (*in vitro*)	Perhaps	Down-regulated	Kishimoto et al. (1996)
DNA breaks, micronuclei	Exposed *in vivo*	Mussels (*Mytilus galloprovincialis*)	No	Increased	Bolognesi et al. (1999)
DNA breaks		Loon		Increased	Evers et al. (2003)
DNA damage	Cultured fibroblasts, cultured neurons	Hamster	N/A	Increased	Costa et al. (1991)
delta-Aminolevulinic acid	Exposed *in situ*	Oyster (*Ostrea edulis*)		Increased	Brock (1993)
DNA synthesis	Immune	Teleost fish	Yes	Suppressed	Burnett (1997)
Nervous tissue/cells					
Monoamine oxidase activity	Brain	Rat	No	Decreased	Chakrabarti et al. (1998)
Serum antobodies to 5 nervous tissue proteins	Serum	Rat	Yes	Increased	El-Fawal et al. (1996)
Muscarinic receptor binding	Brain	Mink	No	Increased	Basu et al. (2006)
Dopaminergic receptor binding				Decreased	Basu et al. (2005b)
Intracellular Ca^{2+} concentration, apoptosis	Cerebellar granule cells	Rat	No	Increased	Rossi et al. (1997)
Cytosolic phospholipase A2 expression → arachadonic acid release	Cultured hippocampal neurons	Mouse	No	Increased	Shanker et al. (2002)

TABLE 5.2 (continued)
Mercury endpoints that may be useful as biomarkers

Endpoint	Tissue or cell	Species	Nonlethal?	Effect	Ref.
Cystine uptake	Cultured astrocytes	Mouse	No	Inhibited	Allen et al. (2002)
Glutamate uptake	Cultured astrocytes	Mouse	No	Inhibited	Aschner et al. (2000)
Other					
Microtubule depolymerazation	Cultured neural cells	Mouse	N/A	Increased, inhibiting cell growth	Miura et al. (1999)
Ocular teratogenesis (microthalmia)	Optic organs	CD-1 mouse embryos	No	Increased	Nemeth et al. (2002)
Ocular development regulatory gene function				Suppressed	
Visual impairment	Optic organs	Great egret	?	Increased	Loerzel et al. (1996)
Visual impairment	Optic organs	Double-crested cormorant	?	Increased	Loerzel et al. (1999)
16s rRNA	Embryo	p53-deficient mouse		Decreased	Knudsen et al. (2000)
Retinol and retinyl palmitate	Liver	Common eider	No	Increased	Wayland et al. (2003)
Retinol/ retinyl palmitate ratio				Increased	Wayland et al. (2003)
Plasma/liver retinol ratio				Decreased	Wayland et al. (2003)
Myosin ATPase activity	Myocardium	Rat	No	Inhibited	Moreira (2003), Moreira et al. (2003)
Morphological asymmetry	Feathers	Loon		Increased	Evers (2004)
Micronuclei	Skin fibroblasts	Beluga whales	Yes	Induced	
Testicular atrophy	Testes	Walleye	No	Increased	Friedmann et al. (1996)
Glycogen storing	Liver	Yellow perch	No	Increased	Hontela et al. (1995)

molecular, electrophysiological, biochemical, and digital imaging are being employed to define the action of mercury on specific nervous cell types and ion channels. Other ongoing research addresses the developmental impacts of *in utero* exposure to mercury, such as its role in causing high-frequency hearing impairment, slowing the rate of information processing, and causing impaired spatial memory. Investigations into the mechanism by which MeHg inhibits cell proliferation may yield changes in MeHg-induced cell cycle regulatory proteins such as p21, GADD 45, and GADD 153, which can be measured in nonlethally-derived tissues by RT-PCR.

5.8 ELEMENTS OF A BIOMONITORING FRAMEWORK

There has been increasing attention in the literature to concerns surrounding the design and evaluation of environmental monitoring programs (e.g., Olsen et al. 1999; Vos et al. 2000). Components discussed by Vos et al. (2000) that would cut across the various aspects of a mercury monitoring network include identifying monitoring objectives, objects, and variables; sampling strategy; data collection; data handling, maintenance; and organization. (Additional general concerns regarding the establishment of a monitoring network that would also be applicable to wildlife monitoring are discussed in Chapter 6.)

While there have been numerous environmental monitoring programs developed over the past several decades (for example, see Olsen et al. 1999) for review of U.S. programs), there have been relatively few efforts that have included systematic monitoring of mercury in wildlife.

5.8.1 MONITORING DESIGN CONSIDERATIONS

In designing a mercury monitoring network that includes a wildlife component, a principal objective would be to document changes in mercury exposure (and potentially effects) relative to changes in mercury loadings to an ecosystem. More specific objectives might include the ability to:

1) Discern spatial differences and temporal changes in mercury exposures (i.e., an assessment of concentrations in wildlife, as well as, potentially, effects) in wide-ranging individual species.
2) Discern temporal changes in exposures (and potentially effects) in more limited range species in specific locations of interest.
3) Identify the role of loadings versus other factors (e.g., food web changes) in mercury exposure changes;
4) Identify the impact of changes in mercury releases from specific sectors and/or geographic areas on exposures in specific locations.

The development of such a network must take into account the numerous variables and factors affecting mercury exposure and effects in wildlife species addressed in this chapter, including geographic and habitat differences (both the broader differences between the 4 major habitat types considered — terrestrial, freshwater, coastal/estuarine, and marine — and differences (including trophic changes) within each habitat type, and host factors (including species migratory status, home range size, functional niche, diet, sex, and age). A mercury monitoring network for wildlife must be integrated into the broader mercury monitoring context, taking into account sources and atmospheric transport, deposition and watershed transport (Chapter 2), cycling within a water body (Chapter 3), and uptake within the aquatic food web

(Chapter 4). Because wildlife (in particular, top-level predators) are the integrators of a mercury signal that has already been integrated across several media (air, water, sediments, and fish), the fourth objective is most challenging, and will require careful attention in overall network development.

A number of medium- to long-term environmental monitoring networks and programs have been developed over the past 4 decades in the United States, all of which have had to deal with network design issues noted here. Some of these efforts are summarized in Table 5.3. As indicated in Table 5.3, some programs have emphasized random or partially random designs, whereas others have utilized nonrandom designs but based on specific site criteria.

Issues of sampling and analysis protocol development have arisen in environmental specimen banking (ESB), which has been the subject of 2 international conferences in the past 15 years (e.g., *Science of the Total Environment*, Vol. 139/140, November 1993; *Chemosphere*, Vol. 34 (9–10), May 1997). Several programs using the ESB approach in North America, and including mercury in the suite of chemicals analyzed, include the Alaska Marine Mammal Tissue Archival Project (AMMTAP) (including analysis of beluga whale, and ringed, bearded, and harbor seal tissue) (Becker et al. 1997); the Marine Mammal Health and Stranding Response Program (MMHSRP) (initially involving sampling of marine mammals outside Alaska, and now incorporating AMMTAP within it) (Becker et al. 1997; Becker 2000); and the Seabird Tissue Archival and Monitoring Project (STAMP) (a newer program involving sampling of common and thick-billed murre and black-legged kittiwake eggs from 12 colonies in Alaska) (Christopher et al. 2002). Other ESB and related contaminant monitoring programs for wildlife have been developed elsewhere, including in Canada, Sweden, Norway, Finland, and Germany (see *Chemosphere*, Vol. 34 (9–10), May 1997). These existing ESB programs and other long-term research and monitoring efforts can potentially assist in the development of a broader mercury biomonitoring effort through both experience in sampling design and storage protocols as well as through the availability of tissue data (and potentially tissue samples) for some species and regions that may be of interest in the development of a new monitoring network.

A program of particular relevance to a continental mercury monitoring network is the Global Loon Mercury Monitoring and Research Cooperative (GLMMRC), which is a long-term monitoring program for levels of mercury in North American freshwater bird species begun in 1991 (Evers 2004). A related but broader effort is the Northeastern Ecosystem Research Cooperative (NERC), formed in 2000 with a goal to promote collaboration among scientists in the northeastern United States and Canada on landscape-level environmental issues (Evers and Clair 2005). Part of this effort has entailed compiling and interpreting a large data set of more than 4700 records of avian mercury levels in northeastern North America. Based on consideration of a number of the factors known to affect mercury levels, the researchers identified 16 bird species that would be suitable as bioindicators for various habitats (Evers et al. 2005). An additional model long-term trend program for wildlife contaminants is the Canadian Wildlife Service's Great Lakes Herring Gull Monitoring Program, which has involved annual collection and analysis of eggs (including for mercury) for the past 3 decades (Pekarik et al. 1998; Hebert et al. 1999).

TABLE 5.3
Selected U.S. environmental monitoring programs involving water quality or wildlife

Program	Agency	Resource sampled	Design features	Sampling frequency
Biomonitoring Status and Trends Program — Fish	FWS	Fish in rivers, Great Lakes	Spatially distributed using site criteria	Every 2 years
Biomonitoring Status and Trends Program — Starlings	FWS	Starlings in terrestrial, riverine habitats	Stratified random	Every 3 years
Environmental Monitoring and Assessment Program — Estuaries	USEPA	Coastal waters (including fish, benthos)	Systematic grid with random start	Annually, with 4-year alternating panels
Environmental Monitoring and Assessment Program — Surface Waters	USEPA	Streams, lakes (including fish)	Systematic grid with random start; 2-phase unequal probability sampling	Annually, with 4-year alternating panels
National Status and Trends	NOAA	Estuaries, coastal waters, Great Lakes (including sediments, mussels, fish)	Spatially distributed using site criteria	Annual
North American Breeding Bird Survey	NBS/CWS	Birds (including in Alaska and portions of Canada and Mexico)	Stratified random (with access restrictions)	Annual
National Stream Quality Accounting Network	USGS	Major rivers	Site criteria	Annual
National Water Quality Assessment Program	USGS	Rivers, streams, aquifers	Priority basin selection; site criteria within basin	9-year cycles; 20 basins every 3 years

(Adapted from Olsen et al. 1999). Abbreviations: CWS: Canadian Wildlife Service; USEPA: US Environmental Protection Agency; FWS: Fish and Wildlife Service; NBS: National Biological Service; NOAA: National Oceanic and Atmospheric Administration; USGS: U.S. Geological Survey.

Assuming limited resources, the wildlife component of a mercury monitoring network would likely have to rely on analyses of tissue from a small subset of candidate species identified earlier in this chapter, so identification of key species that both meet the criteria as good current indicators (see Section 5.5) as well as addressing some of the more specific objectives above would be necessary. Additional criteria would include species that have a strong aquatic link, small home ranges, and be ubiquitous across ecosystems.

In addition, a decision would need to be made whether one is targeting bioindicators of exposure or of effect and/or susceptibility. In the latter case, the ecoepidemiological approach advocated by Gilbertson (1997) could be pursued in the goal of inferring causality, with 7 criteria to be used in demonstrating a causal link, including strength, consistency, and specificity of association, time order, biological gradient, experimental evidence, and biological plausibility (Adams 2003).

5.8.2 Trend Detection: The Florida Everglades Case Study

With the finding of high levels of mercury in Everglades fish and wildlife in 1989, Florida began science, policy, and control initiatives to understand, and, it was hoped, ameliorate, this problem (Figure 5.3). How to do so was all the more mystifying, given the remote and seemingly pristine character of the Everglades. A key element was a program of monitoring, modeling, and research into the sources and cycling of mercury. The importance of trend information was recognized at the outset, including retrospective information and prospective monitoring to assess the efficacy of control policies.

In 1993, Florida began mercury pollution prevention efforts, adopting model legislation to minimize mercury uses in commerce, followed in 1994 by the first emissions limiting rule on mercury emissions from solid waste incinerators (the forerunner of the USEPA Incinerator Maximum Achievable Control Technology (MACT) standard), and in 1995 by adopting the EPA MACT for medical waste incineration. Together, these actions greatly reduced the high concentrations of reactive gas-phase mercury (RGM) from incineration sources, the form that tends

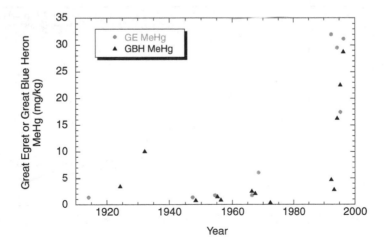

FIGURE 5.3 Increases in mercury concentrations of feathers of great egrets and great blue herons in Florida.

to deposit on a local scale. Further pollution-prevention activities focused on minimizing mercury uses in commercial, industrial, and hospital settings, realizing additional reductions in mercury use, and tighter controls on releases.

5.8.2.1 Retrospective Studies

To understand the antecedents of this problem, however, the first task was to put contemporary data in historical perspective. Had mercury burdens in fish increased, or were the high levels in bass from the Everglades a natural mercury condition of the soils of the area? Investigations elsewhere had found atmospheric deposition to have increased 3-fold at remote sites since the Industrial Revolution, and 5- to 10-fold near cities or industrial centers. Analysis of Everglades sediment cores showed that mercury accumulation rates at the surface were approximately 5-fold higher than in 1900 (Rood et al. 1995), indicating that Everglades sediments reflect the increasing flux of mercury in the environment.

Monitoring mercury in the atmosphere has not been conducted long enough in Florida or the United States to see unambiguous trends, but European studies have documented increasing atmospheric mercury through the 1980s but trending downward for the past decade. This is a hopeful sign that decreasing use of mercury and control of emissions have begun to show the desired effect. How long it will take to detect and evaluate trends in Florida waters and water bodies is not known.

5.8.2.2 Prospective Studies

Upon the initial findings of high levels of mercury in largemouth bass from the Everglades, Florida began trend monitoring of this species that continues to the present. Standard protocols were developed for sampling, testing, and normalizing samples collected from 2 Everglades canal and marsh sites (Figure 5.4).

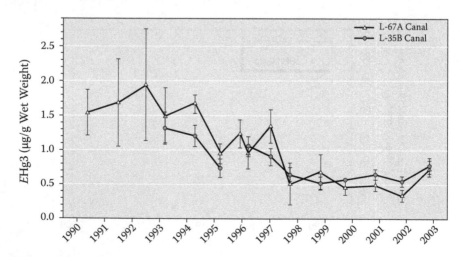

FIGURE 5.4 Mercury concentrations in axial muscle tissue of largemouth bass from 2 canals in the Florida Everglades.

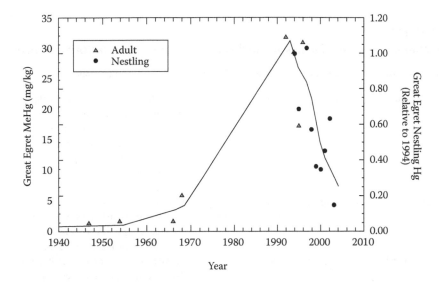

FIGURE 5.5 Reconstruction of time course of great egret feather mercury concentrations as a result of mercury emissions changes in southern Florida, from 1940 through 2003. (*Source:* From Frederick et al. 2004.)

Atmospheric deposition measurements (Guentzel et al. 1995, 1998) in comparison with surface water inputs indicated that deposition flux dominated the mercury budget of the Everglades by about 50:1. Direct measurement of deposition reductions was not possible as deposition collection did not commence until 1992. However, in conjunction with soil accumulation rates, trends in several matrices support the view that mercury deposition increased steadily throughout the 20th century, at an increasing rate post-World War II, peaking around 1991, and declined thereafter, at least in part as a consequence of general reduction of mercury use in the developed world with more accentuated declines in southern Florida.

Collectively, these pollution prevention and control policies have minimized the local-scale problem (hot spot) of mercury emissions and deposition from sources within south Florida, resulting in substantial declines (ca. 80%) in mercury in Everglades fish and wildlife (Figure 5.5). Additional factors have also influenced the bioaccumulation of mercury in Everglades fish and wildlife (including changing sulfate levels), thus indicating the importance of measuring ancillary parameters in assessments of biota response to changing mercury releases, as discussed in previous chapters.

5.8.3 Recommended Wildlife Indicators

Concerning indicator species, as a preliminary assessment in describing the wildlife component of a mercury monitoring network, based on criteria evaluated in Section 5.5 and studies cited herein, additional concerns noted above, and professional judgment of the authors, candidate wildlife species for inclusion in such a network have been selected and are indicated in Table 5.4.

TABLE 5.4
Candidate wildlife species for inclusion in a mercury monitoring network

Habitat type	Species	Age	Tissue	Rationale
Terrestrial	Bicknell's thrush	Adult	Blood	Ability for significant uptake of Hg in montane habitats; high priority for conservation; blood levels related to atmospheric deposition models
	Raccoon	Adult	Hair	Widespread distribution and easily studied; prey for endangered species (Florida panther)
Lake	Common loon	Adult	Blood, egg	Widespread distribution and easily studied; large existing database; sensitivity to Hg; existing database for continental Hg exposure and lab and field effects
Lake/coastal	Herring gull	Adult	Egg	Widespread distribution and easily studied; large existing database documenting trends in Great Lakes
	Bald eagle	Juvenile	Blood	Widespread distribution; large existing database; apparent reproductive effects
	Common tern	Adult	Feather, egg	Widespread distribution and easily studied; large existing database; sensitivity to Hg
Estuarine	Saltmarsh sharp-tailed and seaside sparrows	Adult, juvenile	Blood	Obligate estuarine species; high conservation concern; existing database on exposure; small home range
Riverine	Mink	Adult	Fur[a]	Widespread distribution and easily studied; existing database on exposure and effects; sensitivity to Hg
Marine nearshore	Harbor porpoise[b]	Adult	Muscle[c]	Widespread distribution; some existing data; concern about immunological effects
Marine offshore	Leach's storm-petrel	Adult, juvenile	Blood	Widespread distribution, common, easily sampled at nest site, potential connection with global Hg signal
Comparison across aquatic habitats	Belted kingfisher	Adult, fledged young	Blood, egg	Widespread distribution across habitat types; large and consistent clutch size; ease of capture; relatively high mercury exposures

Notes: a) May require better understanding of relationship with internal tissue mercury concentrations (see discussion in Section 5.6). b) Harbor seal or gray seal would also be candidates for similar reasons, and ringed seal would be appropriate marine mammal target for Arctic biomonitoring, due to wide distribution, significant uptake rates for Hg and importance in diet of polar bears. c) Liver and kidney are more typically analyzed in marine mammal species; but due to both accumulation and detoxification as discussed previously, muscle tissue may be more appropriate indicator, in particular for shorter-term exposures, assuming methodological issues are resolved (see Becker 1991).

Concerning the frequency of sampling, in most cases, sampling could be done annually. However, an important factor to consider in wildlife sampling would be season, as discussed in Section 5.2. Additional ancillary data to be collected on individuals, as previously noted, include whether live or dead specimen, age, sex, and site of collection on tissue.

Most of the species indicated above share key criteria identified earlier (including widespread distribution, relatively easily sampled, sensitivity to changes in mercury availability, and a generally good existing database). As noted earlier in this chapter, numerous wildlife species could potentially serve as bioindicators for mercury exposure and effects. For purposes of this exercise, the authors limited choices to 1 or 2 species per habitat type. As noted previously (Section 5.5), in some cases our assessment found relatively small differences (whether within taxa or by habitat) in suitability of various species to serve as mercury bioindicators. Due to the much larger database on piscivorous wildlife, most priority species we have identified fall within this category. The more recent findings on elevated mercury levels in some passerines, however, indicate the value in incorporating 1 insectivorous species in the network. Additional research on mercury biouptake or sensitivity to methylmercury for individual species, as well as broader considerations of overall network design, may indicate higher value in choosing other species as bioindicators. Assuming resource constraints limited wildlife monitoring to even fewer than the 12 species listed above, at a minimum, we believe 1 species (generally a piscivorous bird) per habitat type (i.e., terrestrial, lake, riverine, and coastal/marine) would be warranted.

Regardless of species chosen in the final design, it is assumed that in most cases sampling would be done annually. In addition, sample collection protocols (including considerations of season/time of year, age, sex, tissue, and sampling location within tissue) would have to be clearly specified.

ACKNOWLEDGMENTS

JB would like to thank Michael Gochfeld for helpful comments on the manuscripts and the following agencies or organizations for funding: USEPA, NIEHS (ESO 5022), DOE through the Consortium for Risk Evaluation with Stakeholder Participation (AI # DE-FC01-95EW55084, DE-FG 26-00NT 40938), New Jersey DEP, and Wildlife Trust. The results, conclusions, and interpretations reported herein are the sole responsibility of the authors, and should not in any way be interpreted as representing the views of the funding agencies.

MFW would like to thank JoAnn Bradley, Miriam Library, California State University – Chico, for heroic document retrieval; and Greg Linder, Jack Campbell, and Glen Lubcke for assistance with reptile and amphibian discussions. Special thanks to Michael Murray and David Evers for manuscript editing.

DE would like to thank Chris DeSorbo for compiling much of the section on the osprey.

MM would like to acknowledge and thank several funding sources that helped support time spent on this project, including the Beldon Fund, the George Gund

Foundation, and the USEPA. Perspectives expressed in this chapter should not be interpreted as representing the views of the funding agencies.

REFERENCES

Abernathy AR, Cumbie PM. 1977. Mercury accumulation by largemouth bass (*Micropterus salmoides*) in recently impounded reservoirs. Bull Environ Contam Toxicol 17:595–602.

Adams S. 2003. Establishing causality between environmental stressors and effects on aquatic environments. Human Environ Risk Assess 9:17–35.

Aihara M, Sharma RP. 1986. Effects of endogenous and exogenous thiols on the distribution of mercurial compounds in mouse tissues. Arch Environ Contam Toxicol 15:629–636.

Albanus L, Frankenberg L, Grant C, von Haartman U, Jernelov A, Nordberg G, Rydalv M, Schutz A, Skerfving S. 1972. Toxicity for cats of methylmercury in contaminated fish from Swedish lakes and of methylmercury hydroxide added to fish. Environ Res 5:425–442.

Allen JW, Shanker G, Aschner M. 2001. Methylmercury inhibits the *in vitro* uptake of the glutathione precursor, cystine, in astrocytes, but not in neurons. Brain Res 894:131–140.

Allen JW, Shanker G, Tan KH, Aschner M. 2002. The consequences of methylmercury exposure on interactive functions between astrocytes and neurons. Neurotoxicology 23:755–759.

Allen P. 1994. Distribution of mercury in the soft tissues of the blue tilapia *Oreochromis aureus* (Steindachner) after acute exposure to mercury II chloride. Bull Environ Contam Toxicol 53:675–683.

Anderson DW, Cahill TMJ, Suchanek TH, Elbert RA. 1997. Relationships between mercury and yearly trends in osprey production and reproductive status at Clear Lake. In: First Annual Clear Lake Science and Management Symposium, September 13, 1997, Proceedings, p. 66–70.

Anderson JR. 1997. Recommendations for the biopsy procedure and assessment of skeletal muscle biopsies. Virchows Arch 431:227–233.

Aoki H. 1970. [Environmental contamination by mercury (Hg series No. 14). 3. Inorganic and organic mercury in human hair and marine fish]. Nippon Eiseigaku Zasshi 24:556–562.

Arnold BS. 2000. Distribution of mercury within different trophic levels of the Okefenokee swamp, within tissues of top level predators, and reproductive effects of methyl mercury in the Nile tilapia (*Oreochromis niloticus*). PhD dissertation, University of Georgia.

Aschner M. 1998. Blood-brain barrier: physiological and functional considerations. In: Handbook of developmental neurotoxicology, p. 339–351.

Aschner M. 1999. Astrocyte response to chemical/physical stress with respect to metallothionein expression. J Neurochem 72:S43.

Aschner M, Aschner JL. 1990. Mercury neurotoxicity: mechanisms of blood-brain barrier transport. Neurosci Biobehav Rev 14:169–176.

Aschner M, Eberle NB, Miller K, Kimelberg HK. 1990. Interactions of methylmercury with rat primary astrocyte cultures: inhibition of rubidium and glutamate uptake and induction of swelling. Brain Res 530:245–250.

Aschner M, Yao CP, Allen JW, Tan KH. 2000. Methylmercury alters glutamate transport in astrocytes. Neurochem Int 37:199–206.

Ashpole SL, Bishop CA, Brooks RJ. 2004. Contaminant residues in snapping turtle (*Chelydra s. serpentina*) eggs from the Great Lakes-St. Lawrence River basin (1999 to 2000). Arch Environ Contam Toxicol 47:240–252.

Atwell L, Hobson KA, Welch HE. 1998. Biomagnification and bioaccumulation of mercury in an arctic marine food web: insights from stable nitrogen isotope analysis. Can J Fish Aquat Sci 55:1114–1121.

Aulerich RJ, Ringer RK, Iwamoto S. 1973. Reproductive failure and mortality in mink fed on Great Lakes fish. J Reprod Fertil Suppl 19:S365–376.

Aulerich RJ, Ringer RK, Iwamoto S. 1974. Effects of dietary mercury on mink. Arch Environ Contam Toxicol 2:43–51.

Bailar JC III. 1995. The practice of meta-analysis. J Clin Epidemiol 48:149–157.

Baker RH. 1983. Michigan mammals. East Lansing (MI): Michigan State University Press.

Balthrop JE, Braddon SA. 1985. Effects of selenium and methylmercury upon glutathione and glutathione-S-transferase in mice. Arch Environ Contam Toxicol 14:197–202.

Baron LA, Ashwood TL, Sample BE, Welsh C. 1997. Monitoring bioaccumulation of contaminants in the belted kingfisher (*Ceryle alcyon*). Environ Monit Assess 47:153–165.

Barr JF. 1986. Population dynamics of the Common Loon (*Gavia immer*) associated with mercury-contaminated waters in northwestern Ontario. Canadian Wildlife Service Occasional Paper 56.

Basu N, Stamler CJ, Loua KM, Chan HM. 2005a. An interspecies comparison of mercury inhibition on muscarinic acetylcholine receptor binding in the cerebral cortex and cerebellum. Toxicol Appl Pharmacol 205:71–76.

Basu N, Klenavic K, Gamberg M, O'Brien M, Evans D, Scheuhammer AM, Chan HM. 2005b. Effects of Mercury on Neurochemical Receptor-Binding Characteristics in Wild Mink. Environ Toxicol Chem 24:1444–1450.

Basu N, Scheuhammer AM, Rouvinen-Watt K, Grochowina N, Klenavic K, Evans RD, Chan HM. 2006. Methylmercury impairs components of the cholinergic system in captive mink (*Mustela vison*). Toxicol Sci 91:202–209.

Batel R, Bihari N, Rinkevich B, Dapper J, Schaecke H, Schroeder HC, Mueller WE. 1993. Modulation of organotin-induced apoptosis by water pollutant methyl mercury in a human lymphoblastoid tumor cell line and a marine sponge. Mar Ecol Progr Ser 93:245–251.

Becker PH. 1992. Egg mercury levels decline with the laying sequence in charadriiformes. Bull Environ Contam Toxicol 48:762–767.

Becker PH, Schuhmann S, Koepff C. 1993. Hatching failure in common terns (*Sterna hirundo*) in relation to environmental chemicals. Environ Pollut 79:207–213.

Becker PR, Wise SA, Koster BJ, Zeisler R. 1991. Alaska Marine Mammal Tissue Archival Project: Revised Collection Protocol. U.S. Department of Commerce, National Institute of Standards and Technology, NISTIR 4529, 33 pp.

Becker PH, Sommer U. 1998. Current contamination of Common Terns (*Sterna hirundo*) with environmental chemicals in Central Europe. Vogelwelt 119:243–249.

Becker PR. 2000. Concentration of chlorinated hydrocarbons and heavy metals in Alaska arctic marine mammals. Mar Pollut Bull 40:819–829.

Becker PR, Wise SA, Thorsteinson L, Koster BJ, Rowles T. 1997. Specimen banking of marine organisms in the United States: current status and long-term prospective. Chemosphere 34:1889–1906.

Bellas J, Vazquez E, Beiras R. 2001. Toxicity of Hg, Cu, Cd, and Cr on early developmental stages of *Ciona intestinalis* (Chordata, Ascidiacea) with potential application in marine water quality assessment. Water Res 35:2905–2912.

Berg W, Johnels A, Jostrand BS, Westermark T. 1966. Mercury content in feathers of Swedish birds from the past 100 years. Oikos 17:71–83.

Birge WJ, Just JJ. 1973. Sensitivity of vertebrate embryos to heavy metals as a criterion of water quality. NTIS Pb Report (Pb-226 850):20 p, 1973 Tax — *Carassius Auratus Tax — Rana Pipiens Tax — Gallus Domesticus,* White Leghorn Tax — *Rana Catesbeiana.*

Birke G, Johnels AG, Plantin LO, Sjostrand B, Skerfving S, Westermark T. 1972. Studies on humans exposed to methyl mercury through fish consumption. Arch Environ Health 25:77–91.

Bishop CA, Koster MD, Chek AA, Hussell DJT, Jock K. 1995a. Chlorinated hydrocarbons and mercury in sediments, red-winged blackbirds (*Agelaius phoeniceus*) and tree swallows (*Tachycineta bicolor*) from wetlands in the Great Lakes–St. Lawrence River basin. Environ Toxicol Chem 14:491–501.

Bishop CA, Lean DR, Brooks RJ, Carey JH, Ng P. 1995b. Chlorinated hydrocarbons in early life stages of the common snapping turtle (*Chelydra serpentina serpentina*) from a coastal wetland on Lake Ontario, Canada. Environ Toxicol Chem 14:421–426.

Bishop CA, Ng P, Pettit KE, Kennedy SW, Stegeman JJ, Norstrom RJ, Brooks RJ. 1998. Environmental contamination and developmental abnormalities in eggs and hatchlings of the common snapping turtle (*Chelydra serpentina serpentina*) from the Great Lakes–St. Lawrence River basin (1989–91). Environ Pollut 101:143–56.

Bishop CA, Rouse JD. 2000. Chlorinated hydrocarbon concentrations in plasma of Lake Erie water snake (*Nerodia sipedon insularum*) and northern water snake (*Nerodia sipedon sipedon*) from the Great Lakes Basin in 1998. Arch Environ Contam Toxicol 39:500–505.

Bjorge A. 2001. How persistent are marine mammal habitats in an ocean of variability? In: Evans PGH, Raga JA, editors, Marine mammals: biology and conservation. New York (NY):Kluwer Academic/Plenum Publishers.

Blakeley BR, Sisoda CS, Mukkar TK. 1980. The effect of methylmercury, tetraethyl lead, and sodium arsenite on the humoral immune response in mice. Toxicol Appl Pharm 52:245–254.

Bleau H, Daniel C, Chevalier G, Van Tra H, Hontela A. 1996. Effects of acute exposure to mercury chloride and methylmercury on plasma cortisol, T3, T4, glucose and liver glycogen in rainbow trout (*Oncorhynchus mykiss*). Aquat Toxicol 34:221–235.

Block E. 1992. Contaminants in great blue heron eggs and young from Dumas Bay and Nisqually heronries, Puget Sound, Washington. U.S. Fish and Wildlife Service OFO-EC93-1.

Blus LJ, Henny CJ, Anderson A, Fitzner RE. 1985. Reproduction, mortality, and heavy metal concentrations in great blue herons from three colonies in Washington and Idaho. Colon Waterbirds 8:110–116.

Bodaly RA, Hecky RE, Fudge RJ. 1984. Increases in fish mercury levels in lakes flooded by the Churchill River diversion, northern Manitoba (Canada). Can J Fish Aquat Sci 41:682–691.

Bolognesi C, Landini E, Roggieri P, Fabbri R, Viarengo A. 1999. Genotoxicity biomarkers in the assessment of heavy metal effects in mussels: experimental studies. Environ Mol Mutagen 33:287–292.

Bonin J, Desgranges JL, Bishop CA, Rodrigue J, Gendron A, Elliott E. 1995. Comparative study of contaminants in the mudpuppy (Amphibia) and the common snapping turtle (Reptilia), St. Lawrence River, Canada. Arch Environ Contam Toxicol 28:184–194.

Bornhausen M, Hagen U. 1984. Operant behavior performance changes in rats after prenatal and postnatal exposure to heavy metals. Ircs Med Sci 12:805–806.

Bouton SN, Frederick PC, Spalding MG, McGill HC. 1999. Effects of chronic, low concentrations of dietary methylmercury on the behavior of juvenile great egrets. Environ Toxicol Chem 18:1934–1939.

Bowerman WW, Evans ED, Giesy P, Postupalsky S. 1994. Using feathers to assess risk of mercury and selenium to bald eagle reproduction in the Great Lakes region. Arch Environ Contam Toxicol 27:294–298.

Bowerman WW, Roe AS, Gilbertson M, Best DA, Sikarskie JG, Summer CL. 2002. Using bald eagles to indicate the health of the Great Lakes environment. Lakes Reservoirs: Res Manage 7:183–187.

Braune BM, Gaskin DE. 1987. Mercury levels in Bonaparte's gulls (*Larus philadelphia*) during autumn molt in the Quoddy Region, New Brunswick, Canada. Arch Environ Contam Toxicol 16:539–550.

Brisbin IL, Jagoe CH, Gaines KR, Cariboldi JC. 1998. Environmental contaminants as concerns for the conservation biology of crocodilians. In Proceedings of the 14th Working Meeting of the Crocodile Specialist Group of the SSC of the IUNC — The World Conservation Union, Gland, Switzerland, p. 155–123.

Britson CA, Threlkeld ST. 1998. Abundance, metamorphosis, developmental, and behavioral abnormalities in *Hyla chrysoscelis* tadpoles following exposure to three agrichemicals and methyl mercury in outdoor mesocosms. Bull Environ Contam Toxicol 61:154–161.

Brock V. 1993. Effects of mercury on physiological conditions and content of the biomarker ALA in the oystery *Ostrea edulis*. Mar Ecol Progr Ser 96:169–175.

Burger J. 1990. Behavioral effects of early postnatal lead exposure in herring gull (*Larus argentatus*) chicks. Pharmacol Biochem Behav 35:7–14.

Burger J. 1993. Metals in avian feathers: bioindicators of environmental pollution. Rev Environ Toxicol 5:203–311.

Burger J. 1994. Heavy metals in avian eggshells: another excretion method. J Toxicol Environ Health 41:207–220.

Burger J. 1995. Heavy metal and selenium levels in feathers of herring gulls: differences due to year, gender, and age at Captree, Long Island. Environ Monit Assess 38:37–50.

Burger J. 2002. Food chain differences affect heavy metals in bird eggs in Barnegat Bay, New Jersey. Environ Res 90:33–39.

Burger J, Gochfeld M. 1988. Metals in tern eggs in a New Jersey estuary: a decade of change. Environ Monit Assess 11:127–135.

Burger J, Gochfeld M. 1991. The common tern: its breeding biology and behavior. New York (NY):Columbia University Press.

Burger J, Gochfeld M. 1993. Heavy metal and selenium levels in feathers of young egrets and herons from Hong Kong and Szechuan, China. Arch Environ Contam Toxicol 25:322–327.

Burger J, Gochfeld M. 1995. Heavy metal and selenium concentrations in eggs of Herring Gulls (*Larus argentatus*): temporal differences from 1989 to 1994. Arch Environ Contam Toxicol 29:192–197.

Burger J, Gochfeld M. 1996a. Family Sternidae (terns). In: DelHoyo J, Elliott AA, Sargatal J, editors, Handbook of the birds of the world, Vol 3. Barcelona: Lynx Edicions, p. 624–667.

Burger J, Gochfeld M. 1996b. Heavy metal and selenium levels in birds at Agassiz National Wildlife Refuge, Minnesota: food chain differences. Environ Monit Assess 43:267–282.

Burger J, Gochfeld M. 1997. Risk, mercury levels, and birds: relating adverse laboratory effects to field biomonitoring. Environ Res 75:160–172.

Burger J, Gochfeld M. 2000a. Effects of lead on birds (Laridae): a review of laboratory and field studies. J Toxicol Environ Health B Crit Rev 3:59–78.

Burger J, Gochfeld M. 2000b. Metals in albatross feathers from Midway Atoll: influence of species, age, and nest location. Environ Res 82:207–21.

Burger J, Gochfeld M. 2000c. On developing bioindicators for human and ecological health. Environ Monit Assess 66:23–46.

Burger J, Gochfeld M. 2001. Metal levels in feathers of cormorants, flamingos and gulls from the coast of Namibia in southern Africa. Environ Monit Assess 69:195–203.

Burger J, Gochfeld M. 2002. Effects of chemicals and pollution on seabirds. In: Schreiber EA, Burger J, editors, Biology of marine birds. Boca Raton (FL): CRC Press.

Burger J, Gochfeld M. 2003. Spatial and temporal patterns in metal levels in eggs of common terns (*Sterna hirundo*) in New Jersey. Sci Total Environ 311:91–100.

Burger J, Gochfeld M. 2004. Bioindicators for assessing human and ecological health. In: Wiersna B, editor, Environmental monitoring. Boca Raton (FL): CRC Press, p. 541–566.

Burger J, Gochfeld M, Rooney AA, Orlando EF, Woodward AR, Guillette Jr LJ. 2000a. Metals and metalloids in tissues of American alligators in three Florida lakes. Arch Environ Contam Toxicol 38:501–508.

Burger J, Kannan K, Geisy JP, Grue C, Gochfeld M. 2002. Effects of environmental pollutants on avian behavior. In: Dell'Omo G, editor, Behavioral ecotoxicology. New York: John Wiley & Sons, p. 337–375.

Burger J, Lord CG, Yurkow EJ, McGrath L, Gaines KF, Brisbin Jr IL, Gochfeld M. 2000b. Metals and metallothionein in the liver of raccoons: utility for environmental assessment and monitoring. J Toxicol Environ Health A 60:243–261.

Burger J, Nisbet IC, Gochfeld M. 1992. Metal levels in regrown feathers: assessment of contamination on the wintering and breeding grounds in the same individuals. J Toxicol Environ Health 37:363–374.

Burger J, Shukla T, Dixon C, Shukla S, McMahon MJ, Ramos R, Gochfeld M. 2001. Metals in feathers of sooty tern, white tern, gray-backed tern, and brown noddy from islands in the North Pacific. Environ Monit Assess 71:71–89.

Burger J, Snodgrass J. 1998. Heavy metals in bullfrog (*Rana catesbeiana*) tadpoles: effects of depuration before analysis. Environ Toxicol Chem 17:2203–2209.

Burger J, Woolfenden GE, Gochfeld M. 1999. Metal concentrations in the eggs of endangered Florida scrub-jays from central Florida. Arch Environ Contam Toxicol 37:385–388.

Burger J, Campbell KR, Campbell TS, Shukla T, Jeitner C, Gochfeld M. 2005. Use of skin and blood as nonlethal indicators of heavy metal contamination in northern water and snakes (Nerodia sipedon). Arch Environ Contam Toxicol 49:232–238.

Burgess N, Beauchamp S, Brun G, Clair T, Roberts C, Rutherford L, Tordon R, Vaidya O. 1998. Mercury in Atlantic Canada: a progress report. Sackville, New Brunswick (Canada): Environment Canada.

Burgess NM, Braune BM. 2002. Increasing trends in mercury concentrations in Atlantic and Arctic seabird eggs in Canada. SETAC poster.

Burgess NM, Evers DC, Kaplan JD. 2005. Mercury and other contaminants in common loons breeding in Atlantic Canada. Ecotoxicology 14:241–252.

Burnett KG. 1997. Evaluating intracellular signaling pathways as biomarkers for environmental contaminant exposures. Am Zoologist 37:585–594.

Butler RW, Whitehead PE, Breault AM, Moul IE. 1995. Colony effects on fledging success of great blue herons (*Ardea herodias*) in British Columbia. Colon Waterbirds 18:159–165.

Cahill TM, Anderson DW, Elbert RA, Perley BP, Johnson DR. 1998. Elemental profiles in feather samples from a mercury-contaminated lake in central California. Arch Environ Contam Toxicol 35:75–81.

Calabrese A, Thurberg FP, Dawson MA, Wenzloff DR. 1975. Sublethal physiological stress induced by cadmium and mercury in the winter flounder (*Pseudopluronectes americanus*). In: Koeman JH, Strik JJ, editors, Sublethal effects of toxic chemicals on aquatic animals. Amsterdam: Elsevier.

Campbell KR, Campbell TS. 2001. The accumulation and effects of environmental contaminants on snakes: a review. Environ Monitor Assess 70:253–301.

Campbell KR, Ford CJ, Levine DA. 1998. Mercury distribution in Poplar Creek, Oak Ridge, Tennessee, USA. Environ Toxicol Chem 17:1191–1198.

Canli M, Furness RW. 1993. Toxicity of heavy metals dissolved in sea water and influences of sex and size on metal accumulation and tissue distribution in the Norway lobster (*Nephrops norvegicus*). Mar Environ Res 36:217–236.

Caurant F, Navarro M, Amiard JC. 1996. Mercury in pilot whales: possible limits to the detoxification process. Sci Total Environ 186:95–104.

Chakrabarti SK, Loua KM, Bai C, Durham H, Panisset JC. 1998. Modulation of monoamine oxidase activity in different brain regions and platelets following exposure of rats to methylmercury. Neurotoxicol Teratol 20:161–168.

Champoux L. 1996. PCBS, dioxins, and furans in hooded merganser, common merganser, and mink collected along the St. Maurice River, near La Tuque, Quebec. Environ Pollut 92:147–153.

Champoux L, Masse D, Evers DC, Lane O. (In press). Assessment of mercury exposure and potential effects in common loons in Quebec. Hydrobiologia.

Champoux L, Rodrigue J, Desgranges JL, Trudeau S, Hontela A, Boily M, Spear P. 2002. Assessment of contamination and biomarker responses in two species of herons on the St. Lawrence River. Environ Monit Assess 79:193–215.

Chang LW. 1990. The neurotoxicology and pathology of organomercury, organolead, and organotin. J Toxicol Sci 15:125–151.

Chen CY, Stemberger RS, Klaue B, Blum JD, Pickhardt PC, Folt CL. 2000. Accumulation of heavy metals in food web components across a gradient of lakes. Limnol Oceanogr 45:1525–1536.

Christopher SJ, Vander Pol SS, Pugh RS, Day RD, Becker PR. 2002. Determination of mercury in the eggs of Common Murres (*Uria Aalge*) for the Seabird Tissue Archival and Monitoring Project. J Anal Atomic Spectrom 17:780–785.

Cifuentes JM, Becker PH, Sommer U, Pacheco P, Schlatter R. 2003. Seabird eggs as bioindicators of chemical contamination in Chile. Environ Pollut 126:123–137.

Citterio S, Aina R, Labra M, Ghiani A, Fumagalli P, Sgorbati S, Santagostino A. 2002. Soil genotoxicity assessment: a new strategy based on biomolecular tools and plant bioindicators. Environ Sci Technol 36:2748–2753.

Clark Jr DR, Bickham JW, Baker DL, Cowman DF. 2000. Environmental contaminants in Texas, USA, wetland reptiles: evaluation using blood samples. Environ Toxicol Chem 19:2259–2265.

Clark Jr DR, Hill EF, Henry PF. 1991. Comparative sensitivity of little brown bats *Myotis-lucifugus* to acute dosages of sodium cyanide. Bat Res News 32:68.

Clark Jr DR, Hothem RL. 1991. Mammal mortality at Arizona, California, and Nevada gold mines using cyanide extraction. Calif Fish Game 77:61–69.

Clark KE, Stansley W, Niles LJ. 2001. Changes in contaminant levels in New Jersey osprey eggs and prey, 1989 to 1998. Arch Environ Contam Toxicol 40:277–284.

Costa M, Christie NT, Cantoni O, Zelikoff JT, Wang XW, Rossman T. 1991. DNA damage by mercury compounds: an overview. In: Suzuki T, Imura N, Clarkson TW, editors, Advances in mercury toxicology — Rochester Series on Environmental Toxicity. New York (NY): Plenum Press, p. 255–273.

Counard CJ. 2000. Mercury exposure and effects on Common Loon (*Gavia immer*) behavior in the Upper Midwestern United States. University of Minnesota. MS thesis.

Cumbie PM. 1975a. Mercury in hair of bobcats and raccoons. J Wild Manage 39:419–425.

Cumbie PM. 1975b. Mercury levels in Georgia otter, mink, and freshwater fish. Bull Environ Contam Toxicol 14:193–196.

Custer CM, Custer TW, Rosiu CJ, Melancon MJ, Bickham JW, Matso CW. 2005. Exposure and effects of 2,3,7,8-tetrachlorodibenzo-*p*-dioxin in tree swallows (*Tachycineta bicolor*) nesting along the Woonasquatucket River, Rhode Island, USA. Environ Toxicol Chem 24:93–109.

Cuvin-Aralar M, Furness R. 1990. Tissue distribution of mercury and selenium in minnows, *Phoxinus phoxinus*. Bull Environ Contam Toxicol 45:775–782.

Dansereau M, Lariviere N, Du Tremblay D, Belanger D. 1999. Reproductive performance of two generations of female semidomesticated mink fed diets containing organic mercury contaminated freshwater fish. Arch Environ Contam Toxicol 36:221–226.

Das K, Debacker V, Pillet S, Bouquegneau JM. 2003. Heavy metals in marine mammals. In: Vos JG, Bossart GD, Fournier M, O'Shea TJ, editors, Toxicology of marine mammals. London: Taylor & Francis.

Davis LE, Kornfeld M, Mooney HS, Fiedler KJ, Haaland KY, Orrison WW, Cernichiari E, Clarkson TW. 1994. Methylmercury poisoning: long-term clinical, radiological, toxicological, and pathological studies of an affected family. Ann Neurol 35:680–688.

Dawson MA. 1982. Effects of long-term mercury exposure on hematology of striped bass, *Morone-Saxatilis*. Fish Bull 80:389–392.

De Solla SR, Bishop CA, Pettit KE, Elliott JE. 2002. Organochlorine pesticides and polychlorinated biphenyls (PCBs) in eggs of red-legged frogs (*Rana aurora*) and northwestern salamanders (*Ambystoma gracile*) in an agricultural landscape. Chemosphere 46:1027–1032.

Delany M, Bell J, Sundlof S. 1988. Concentrations of contaminants in muscle of the American alligator in Florida. J Wildl Dis 24:62–66.

DesGranges J-L, Rodrigue J, Tardif B, Laperle M. 1998. Mercury accumulation and biomagnification in ospreys (*Pandion haliaetus*) in the James Bay and Hudson Bay regions of Quebec. Arch Environ Contam Toxicol 35:330–341.

DesGranges J-L, Rodrigue J, Tardif B, Laperle M. 1999. In: Lucotte, EA, editor, Breeding success of osprey under high seasonal methylmercury exposure. New York: Springer.

Dickinson PJ, LeCouteur RA, Anderson JR, O'Rourke KS, Ike RW, Pearl GS, Ghatak NR, Braund KG, Edwards RH, Round JM, et al. 2002. Muscle and nerve biopsy: recommendations for the biopsy procedure and assessment of skeletal muscle biopsies. Vet Clin North Am Small Anim Pract 32:63–102.

Dieter MP, Ludke JL. 1975. Studies on the combined effects of organophosphates and heavy metals in birds. I. Plasma and brain cholinesterase in Coturnix quail fed methyl mercury and orally dosed with parathion. Bull Environ Contam Toxicol 13:257–262.

Dock L, Rissanen RL, Vahter M. 1994. Demethylation and placental transfer of methyl mercury in the pregnant hamster. Toxicology 94:1–3.

Doi R. 1991. Individual difference of methylmercury metabolism in animals and its significance in methylmercury toxicity. In: Suzuki T, Imura N, Clarkson TW, editors, Advances in mercury toxicology. New York (NY): Plenum Press.

Doka SE, McNicol DK, Mallory ML, Wong I, Minns CK, Yan ND. 2003. Assessing potential for recovery of biotic richness and indicator species due to changes in acidic deposition and lake pH in five areas of southeastern Canada. Environ Monit Assess 88:53–101.

Duvall SE, Barron MG. 2000. A screening level probabilistic risk assessment of mercury in Florida Everglades food webs. Ecotoxicol Environ Saf 47:298–305.

Eaton RDP, Secord DC, Hewitt P. 1980. An experimental assessment of the toxic potential of mercury in ringed-seal liver for adult laboratory cats. Toxicol Appl Pharmacol 55:514–521.

Echeverria D, Woods JS, Heyer NJ, Rohlman D, Farin FM, Li T, Garabedian CE. 2006. The association between a genetic polymorphism of coproporphyrinogen oxidase, dental mercury exposure and neurobehavioral response in humans. Neurotoxicol Teratol 28:39–48.

Echeverria D, Woods JS, Heyer NJ, Rohlman DS, Farin FM, Bittner Jr AC, Li T, Garabedian C. 2005. Chronic low-level mercury exposure, BDNF polymorphism, and associations with cognitive and motor function. Neurotoxicol Teratol 27:781–796.

Echols KR, Tillitt DE, Nichols JW, Secord AL, McCarty JP. 2004. Accumulation of PCB congeners in nestling tree swallows (*Tachycineta bicolor*) on the Hudson River, New York. Environ Sci Technol 38:6240–6246.

Eisemann JD, Beyer WN, Bennetts RE, Morton A. 1997. Mercury residues in South Florida apple snails (*Pomacea paludosa*). Bull Environ Contam Toxicol 58:739–743.

Eisler R. 1987. Mercury hazards to fish, wildlife and invertebrates: a synoptic review. US Fish and Wildlife Service Contaminant Hazard Reviews 10, 90 pp.

Eisler R. 2006. Mercury: hazards to living organisms. Boca Raton (FL): Taylor & Francis.

El-Fawal HA, Gong Z, Little AR, Evans HL. 1996. Exposure to methyl mercury results in serum autoantibodies to neurotypic and gliotypic proteins. Neurotoxicology 17:267–276.

Elbert RA, Anderson DW. 1998. Mercury levels, reproduction, and hematology of western grebes in three California lakes. Environ Toxicol Chem 17:210–213.

Ellegren H, Lindgren G, Primmer CR, Moller AP. 1997. Fitness loss and germline mutations in barn swallows breeding in Chernobyl. Nature 389:593–596.

Elliott JE, Butler RW, Norstrom RJ, Whitehead PE. 1989. Environmental contaminants and reproductive success of great blue herons, *Ardea herodias*, in British Columbia, Canada 1986–1987. Environ Pollut 59:91–114.

Elliott JE, Machmer MM, Henny CJ, Wilson LK, Norstrom RJ. 1998. Contaminants in ospreys from the Pacific Northwest. I. Trends and patterns in polychlorinated dibenzo-*p*-dioxins and -dibenzofurans in eggs and plasma. Arch Environ Contam Toxicol 35:620–631.

Elliott JE, Machmer MM, Wilson LK, Henny CJ. 2000. Contaminants in ospreys from the Pacific Northwest. II. Organochlorine pesticides, polychlorinated biphenyls, and mercury, 1991–1997. Arch Environ Contam Toxicol 38:93–106.

Ensor KL, Helwig DD, Wemmer LC. 1992. Mercury and lead in Minnesota common loons (*Gavia immer*). Minnesota Pollution Control Agency Water Quality Division.

Eriksson MOG, Johansson I, Ahlgren C. 1992. Levels of mercury in eggs of red-throated diver *Gavia stellata* and black-throated diver *G. arctica* in southwest Sweden. Ornis Svecica 2:29–36.

Escher BI, Hermens JLM. 2002. Modes of action in ecotoxicology: their role in body burdens, species sensitivity, QSARs and mixture effects. Environ Sci Technol 36:4201–4217.

Evans RD. 1993. Mercury and other metals in bald eagle feathers and other tissues from Michigan, nearby areas of Minnesota, Wisconsin, Ohio, Ontario, and Alaska 1985–89. Michigan Department of Environmental Quality, Wildlife Division 3200.

Evans RD. 1995. Mercury levels in otters. Proc 1995 Canadian Mercury Network Workshop. Environment Canada. http://rese.ca/eman/reports/publications/mercury95/part2.html.

Evans RD, Addison EM, Villeneuve JY, MacDonald KS, Joachim DG. 1998. An examination of spatial variation in mercury concentrations in otter (*Lutra canadensis*) in south-central Ontario. Sci Total Environ 213:239–245.

Evans RD, Addison EM, Villeneuve JY, MacDonald KS, Joachim DG. 2000. Distribution of inorganic and ethylmercury among tissues in mink (*Mustela vison*) and otter (*Lutra canadensis*). Environ Res 84:133–139.

Evers D, Lane O, Savoy L, Goodale W. 2004. Assessing the impacts of methylmercury on piscivorous wildlife using a wildlife criterion value based on the common loon, 1998–2003. Biodiversity Research Institute (Report BRI2004-05).

Evers DC. 2004. Status assessment and conservation plan for the Common Loon (*Gavia immer*) in North America. US Fish and Wildlife Service Technical Report.

Evers DC, Burgess N, Major A, Champoux L, Goodale W, Taylor R. 2005. Patterns of mercury exposure in the avian community of northeastern North America. Ecotoxicology 14:193–221.

Evers DC, Clair TA. 2005. Mercury in northeastern North America: a synthesis of existing databases. Ecotoxicology 14:7–14.

Evers DC, Jodice P. 2002. Winter population dynamics of Common Loons on the Florida gulf coast: a preliminary report. Unpublished report, alaska.fws.gov/mbsp/mbm/loons/pdf/Common_Loon_Status_Assessment.pdf.

Evers DC, Kaplan JD, Meyer MW, Reaman PS, Braselton WE, Major A, Burgess N, Scheuhammer AM. 1998. Geographic trends in mercury measured in common loon feathers and blood. Environ Toxicol Chem 17:173–183.

Evers DC, Lane OP, Savoy L. 2003a. Assessing the impacts of methylmercury on piscivorous wildlife using a wildlife criterion value based on the Common Loon, 1998–2002. BioDiversity Research Institute, Falmouth, ME.

Evers DC, Reaman P. 1997. A comparison of mercury exposure between artificial impoundments and natural lakes measured in Common Loons and their prey, sediments, and game fish. BioDiversity Research Institute, Falmouth, ME.

Evers DC, Taylor KM, Major A, Taylor RJ, Poppenga RH, Scheuhammer AM. 2003b. Common loon eggs as indicators of methylmercury availability in North America. Ecotoxicology 12:69–81.

Eyl TB. 1971. Organic-mercury food poisoning. N Engl J Med 284:706–709.

Faber RA, Risebrough RW, Pratt HM. 1972. Organochlorines and mercury in common egrets and great blue herons. Environ Pollut 3:111–122.

Farrington JW. 1991. Biogeochemical processes governing exposure and uptake of organic pollutant compounds in aquatic organisms. Environ Health Perspect 90:75–84.

Farris FF, Dedrick RL. 1993. Absorption of methylmercury from hair ingested by rats. Life Sci 53:1023–1029.

Farris FF, Dedrick RL, Allen PV, Smith JC. 1993. Physiological model for the pharmacokinetics of methyl mercury in the growing rat. Toxicol Appl Pharmacol 119:74–90.

Fenton MB. 1992. Bats. New York (NY): Facts on File.

Fevold BM, Meyer MW, Rasmussen PW, Temple SA. 2003. Bioaccumulation patterns and temporal trends of mercury exposure in Wisconsin common loons. Ecotoxicology 12:83–93.

Fimreite N. 1971. Effects of methylmercury on ring-necked pheasants, with special reference to reproduction. Canadian Wildlife Service, Occasional Paper 9, 37 pp.

Fimreite N, Fyfe RW, Keith JA. 1970. Mercury contamination of Canadian prairie seed eaters and their avian predators. Can Field-Nat 84:269–276.

Fimreite N, Holsworth WN, Keith JA, Pearce PA, Gruchy IM. 1971. Methyl mercury in fish and fish-eating birds from sites of industrial contamination in Canada. Can Field Natural 85:2211–2220.

Fimreite N, Karstad L. 1971. Effects of dietary methyl mercury on red-tailed hawks. J Wildl Manage 35:292–300.

Finley MT, Stendall RC. 1978. Survival and reproductive success of black ducks fed methyl-mercury. Environ Pollut 16:51–64.

Finocchio DV, Luschei ES, Mottet NK, Body RL. 1980. Effects of methylmercury on the visual system of the rhesus macaque (*Macaca mulatta*). I. Pharmacokinetics of chronic methylmercury related to changes in vision and behavior. In: Merigan WH, Weiss B, editors, Neurotoxicity of the visual system, New York: Raven Press, p. 113–122.

Fitzgerald WF, Mason RP. 1996. The global mercury cycle oceanic and anthropogenic aspects. In: Baeyens W, Ebinghaus R, Vasiliev O, editors, Global and regional mercury cycles: sources, fluxes and mass balances, Dordrecht, the Netherlands: Kluwer Academic Publishers, p. 85–108.

Fitzgerald WF, Mason RP. 1997. Biogeochemical cycling of mercury in the marine environment. In: Sigel A, Sigel H, editors, Metal ions in biological systems, Vol. 34, Mercury and its effects on environment and biology. New York (NY): Marcel Dekker, p. 53–110.

Fleming WJ, Pullin BP, Swineford DM. 1985. Population trends and environmental contaminants in herons in the Tennessee Valley USA 1980–1981. Colon Waterbirds 7:63–73.

Florida Department of Environmental Protection. 2003. Integrating atmospheric mercury deposition with aquatic cycling in South Florida: an approach for conducting a total maximum daily load analysis for an atmospherically derived pollutant. Tallahassee, FL: Florida Department of Environmental Protection.

Foley RE, Jackling SJ, Sloan RJ, Brown MK. 1988. Organochlorine and mercury residues in wild mink and otter. Comparison with fish. Environ Toxicol Chem 7:363–374.

Fossi C, Lari L, Mattei N, Savelli C, Sanchez-Hernandez JC, Castellani S, Depledge M, Bamber S, Walker C, Savaa D, et al. 1996. Biochemical and genotoxic biomarkers in the Mediterranean crab (*Carcinus aestuarii*) experimentally exposed to polychlorobiphenyls, benzopyrene and methyl-mercury. Mar Environ Res 42:29–32.

Fournier F, Karasov WH, Kenow KP, Meyer MW, Hines RK. 2002. The oral bioavailability and toxicokinetics of methylmercury in common loon (*Gavia immer*) chicks. Comp Biochem Physiol A Mol Integr Physiol 133:703–714.

Fox DA, Boyes WK. 2001. Toxic responses of the ocular and visual system. In: Klaassen CD, editor, Casarett & Doull's toxicology: the basic science of poisons, 6th ed. New York (NY): McGraw-Hill, p. 565–595.

Fox GA. 1976. Eggshell quality: its ecological and physiological significance in a DDE-contaminated common tern population. Wilson Bull 88:459–477.

Fox GA. 1992. Epidemiological and pathobiological evidence of contaminant-induced alterations in sexual development in free-living wildlife. In: Colborn T, Clement C, editors, Chemically-induced alterations in sexual and functional development: the wildlife/human connection. Princeton (NJ): Princeton Scientific Publishing Company, p. 147–158.

Fox GA, Yonge KS, Sealy SG. 1980. Breeding performance, pollutant burden, and eggshell thinning in Common Loons *Gavia immer* nesting on a boreal forest lake. Ornis Scand 11:243–248.

Frederick PC, Hylton B, Heath JA, Spalding MG. 2004. A historical record of mercury contamination in southern Florida (USA) as inferred from avian feather tissue. Environ Toxicol Chem 23:1474–1478.

Frederick PC, Spalding MG, Dusek R. 2002. Wading birds as bioindicators of mercury contamination in Florida, USA: annual and geographic variation. Environ Toxicol Chem 21:163–167.

Frederick PC, Spalding MG, Sepulveda MS, Williams G, Nico L, Robins R. 1999. Exposure of Great egret (*Ardea albus*) nestlings to mercury through diet in the Everglades ecosystem. Arch Environ Contamin Toxicol Chem 18:1940–1947.

Friedmann AS, Watzin MC, Brinck-Johnsen T, Leiter JC. 1996. Low levels of dietary methyl-mercury inhibit growth and gonadal development in juvenile walleye (*Stizostedion vitreum*). Aquat Toxicol 35:265–278.

Furness RW, Camphuysen K. 1997. Seabirds as monitors of the marine environment. Ices J Mar Sci 54:726–737.

Furness RW, Lewis SA, Mills JA. 1990. Mercury levels in the plumage of red-billed gulls (*Larus novaehollandiae scopulinus*) of known sex and age. Environ Pollut 63:33–39.

Furness RW, Muirhead SJ, Woodburn M. 1986. Using bird feathers to measure mercury in the environment: relationship between mercury content and moult. Mar Pollut Bull 17:27–30.

Gaines KF, Romanek CS, Boring CS, Lord CG, Gochfeld M, Burger J. 2002. Using raccoons as an indicator species for metal accumulation across trophic levels — a stable isotope approach. J Wildl Manage 66:811–821.

Gawel JE, Trick CG, Morel FM. 2001. Phytochelatins are bioindicators of atmospheric metal exposure via direct foliar uptake in trees near Sudbury, Ontario, Canada. Environ Sci Technol 35:2108–2113.

Gilbertson M. 1974. Seasonal changes in organochlorine compounds and mercury in common terns of Hamilton Harbour, Ontario. Bull Environ Contam Toxicol 12:726–732.

Gilbertson M. 1997. Great Lakes forensic toxicology and the implications for research and regulatory programs. Environ Toxicol Chem 16:1771–1778.

Gilbertson M, Kubiak TJ, Ludwig JP, Fox G. 1991. Great Lakes Embryo Mortality, Edema and Deformity Syndrome (GLEMEDS) in colonial fish-eating birds: similarity to chick edema disease. J Toxicol Environ Health 33:455–520.

Gochfeld M. 1980. Tissue distribution of mercury in normal and abnormal young common terns. Mar Pollut Bull 11:362–366.

Gochfeld M. 1997. Spatial patterns in a bioindicator: heavy metal and selenium concentrations in eggs of herring gulls (*Larus argentatus*) in the New York Bight. Arch Environ Contam Toxicol 33:63–70.

Gochfeld M, Burger J. 1998. Temporal trends in metal levels in eggs of the endangered roseate tern (*Sterna dougallii*) in New York. Environ Res 77:36–42.

Goering PL, Fisher BR, Chaudhary PP, Dick CA. 1992. Relationship between stress protein induction in rat kidney by mercuric chloride and nephrotoxicity. Toxicol Appl Pharmacol 113:184–191.

Golet WJ, Haines TA. 2001. Snapping turtles (*Chelydra serpentina*) as monitors for mercury contamination of aquatic environments. Environ Monit Assess 71:211–220.

Grandjean P, Weihe P, White RF, Debes F. 1998. Cognitive performance of children prenatally exposed to "safe" levels of methylmercury. Environ Res 77:165–172.

Grandjean P, Weihe P, White RF, Debes F, Araki S, Yokoyama K, Murata K, Sorensen N, Dahl R, Jorgensen PJ. 1997. Cognitive deficit in 7-year-old children with prenatal exposure to methylmercury. Neurotoxicol Teratol 19:417–428.

Grasman KA, Scanlon PF, Fox GA. 1998. Reproductive and physiological effects of environmental contaminants in fish-eating birds of the Great Lakes: a review of historical trends. Environ Monit Assess 53:117–145.

Grue CE, O'Shea TJ, Hoffman DJ. 1984. Lead concentrations and reproduction in highway-nesting barn swallows (*Hirundo rustica*). Condor 86:383–389.

Guentzel JL, Landing WM, Gill GA, Pollman CD. 1995. Atmospheric deposition of mercury in Florida: the FAMS Project (1992–1994). Water, Air Soil Pollut 80:393–402.

Guentzel JL, Landing WM, Gill GA, Pollman CD. 1998. Mercury and major ions in rainfall, throughfall, and foliage from the Florida Everglades. Sci Total Environ 213:43–51.

Guillette LJJ, Crain DA, Rooney A. 1994. Endocrine-disrupting environmental contaminants and reproductive abnormalities in reptiles. Comments Toxicol 5:381–399.

Guillette LJJ, Pickford DB, Crain DA, Rooney AA, Percival HF. 1996. Reduction in penis size and plasma testosterone concentrations in juvenile alligators living in a contaminated environment. Gen Compar Endocrinol 101:32–42.

Hart CA, Nisbet IC, Kennedy SW, Hahn ME. 2003. Gonadal feminization and halogenated environmental contaminants in Common Terns (*Sterna hirundo*): evidence that ovotestes in male embryos do not persist to the prefledging stage. Ecotoxicol Environ Saf 12:125–140.

Hays H, Risebrough R. 1972. Pollutant concentrations in abnormal young terns from Long Island Sound. The Auk 89:19–35.

Heaton-Jones TG, Homer BL, Heaton-Jones DL, Sundlof SF. 1997. Mercury distribution in American alligators (*Alligator mississippiensis*) in Florida. J Zoo Wildl Med 28:62–70.

Hebert C, Nordstrom R, Weseloh DV. 1999. A quarter century of environmental surveillance: the Canadian Wildlife Service's Great Lakes Herring Gull Monitoring Program. Environ Rev 7:147–166.

Heinz G. 1975. Effects of methylmercury on approach and avoidance behavior of mallard ducklings. Bull Environ Contam Toxicol 13:554–564.

Heinz G. 1976. Methylmercury: second-year feeding effects on mallard reproduction and duckling behavior. J Wildl Manage 40:82–90.

Heinz G. 1996. Mercury poisoning in wildlife. In: Fairbrother A, Locke LN, Hoff GL, editors, Non infectious diseases of wildlife. Ames (IA): Iowa State University Press, Wildlife Disease Association.

Heinz GH. 1979. Methylmercury: reproductive and behavioral effects on three generations of mallard ducks. J Wildl Manage 43:394–401.

Heinz GH, Hoffman DJ. 1998. Methylmercury chloride and selenomethionine interactions on health and reproduction in mallards. Environ Toxicol Chem 17:139–145.

Heinz GH, Percival HF, Jennings ML. 1991. Contaminants in American alligator eggs from Lake Apopka, Lake Griffin, and Lake Okeechobee, Florida. Environ Monit Assess 16:277–285.

Henny CJ, Blus LJ, Thompson SP, Wilson UW. 1989. Environmental contaminants, human disturbances and nesting of double-crested cormorants in northwestern Washington (USA). Colon Waterbirds 12:198–206.

Henny CJ, Hill EF, Hoffman DJ, Spalding MG, Grove RA. 2002. Nineteenth century mercury: hazard to wading birds and cormorants of the Carson River, Nevada. Ecotoxicology 11:213–231.

Henny CJ, Kaiser JL, Grove RA, Bentley VR, Elliott JE. 2003. Biomagnification factors (fish to osprey eggs from Willamette River, Oregon, U.S.A.) for PCDDs, PCDFs, PCBs and OC pesticides. Environ Monit Assess 84:275–315.

Heo Y, Lee WT, Lawrence DA. 1997. *In vivo* the environmental pollutants lead and mercury induce oligoclonal T cell responses skewed toward type-2 reactivities. Cell Immunol 179:185–195.

Heyer NJ, Bittner Jr AC, Echeverria D, Woods JS. 2006. A cascade analysis of the interaction of mercury and coproporphyrinogen oxidase (CPOX) polymorphism on the heme biosynthetic pathway and porphyrin production. Toxicol Lett 161:159–166.

Hilmy AM, El Domiaty NA, Daabees AY, Moussa FI. 1987. Short-term effects of mercury on survival, behaviour, bioaccumulation and ionic pattern in the catfish (*Clarias lazera*). Comp Biochem Physiol C 87:303–308.

Hindell MA, Brothers N, Gales R. 1999. Mercury and cadmium concentrations in the tissues of three species of southern albatrosses. Polar Biol 22:102–108.

Hoffman DJ, Heinz GH. 1998. Effects of mercury and selenium on glutathione metabolism and oxidative stress in mallard ducks. Environ Toxicol Chem 17:161–166.

Hoffman DJ, Ohlendorf HM, Marn CM, Pendleton GW. 1998. Association of mercury and selenium with altered glutathione metabolism and oxidative stress in diving ducks from the San Francisco Bay region, USA. Environ Toxicol Chem 17:167–172.

Hoffman DJ, Spalding MG, Frederick PC. 2005. Subchronic effects of methylmercury on plasma and organ biochemistries in great egret nestlings. Environ Toxicol Chem 24:3078–3084.

Hoffman RD. 1980. Total mercury in heron and egret eggs and excreta. Ohio J Sci 80:43–45.

Holloway J, Scheuhammer AM, Chan HM. 2003. Assessment of white blood cell phagocytosis as an immunological indicator of methylmercury exposure in birds. Arch Environ Contam Toxicol 44:493–501.

Honda K, Nasu T, Tatsukawa R. 1986. Seasonal changes in mercury accumulation in the black-eared kite, (*Milvus migrans lineatus*). Environ Pollut 42:325–334.

Hontela A, Dumont P, Duclos D, Fortin R. 1995. Endocrine and metabolic dysfunction in yellow perch, *Perca flavescens*, exposed to organic contaminants and heavy metals in the St. Lawrence River. Environ Toxicol Chem 14:725–731.

Hothem RL, Roster DL, King KA, Keldsen TK, Marois KC, Wainwright SE. 1995. Spatial and temporal trends of contaminants in eggs of wading birds from San Francisco Bay, California. Environ Toxicol Chem 14:1319–1331.

Houpt KA, Essick LA, Shaw EB, Alo DK, Gilmartin JE, Gutenmann WH, Littman CB, Lisk DJ. 1988. A tuna fish diet influences cat behavior. J Toxicol Environ Health 24:161–172.

Huckabee JW, Cartam FO, Kennington GS, Camenzind FJ. 1973. Mercury concentration in the hair of coyotes and rodents in Jackson Hole, Wyoming. Bull Environ Contam Toxicol 9:37–43.

Hughes KD, Ewins PJ, Clark KE. 1997. A comparison of mercury levels in feathers and eggs of osprey (*Pandion haliaetus*) in the North American Great Lakes. Arch Environ Contam Toxicol 33:441–452.

Ilback NG, Sundberg J, Oskarsson A. 1991. Methyl mercury exposure via placenta and milk impairs natural killer (NK) cell function in newborn rats. Toxicol Lett 58:149–158.

InSug O, Datar S, Koch CJ, Shapiro IM, Shenker BJ. 1997. Mercuric compounds inhibit human monocyte function by inducing apoptosis: evidence for formation of reactive oxygen species, development of mitochondrial membrane permeability transition and loss of reductive reserve. Toxicology 124:211–224.

Ioannidis JP, Lau J. 1999. Pooling research results: benefits and limitations of meta-analysis. Jt Comm J Qual Improv 25:462–469.

Jaffe D, Prestbob E, Swartzendrubera P, Weiss-Penziasa P, Katoc S, Takamid A, Hatakeyamad S, Kajiic Y. 2005. Export of atmospheric mercury from Asia. Atmos Environ 39:3029–3038.

Jarvinen AW, Ankley G. 1999. Linkage of effects to tissue residues: development of a comprehensive database for aquatic organisms exposed to inorganic and organic chemicals. Pensacola (FL): SETAC.

Jenkins DW. 1980. Biological monitoring of toxic trace metals. Vol 1, Biological monitoring and surveillance. U.S. Environmental Protection Agency 600/3-80-091.

Jenkins DW. 1981. Biological monitoring of toxic trace elements, Vol 1, Biological monitoring and surveillance; Vol 2, Toxic trace metals in plants and animals of the world, Parts I, II, and III. U.S. Environmental Protection Agency, Environmental Systems Laboratory EPA-600/S3-80-090.

Kelly CA, Rudd JWM, St. Louis VL, Heyes A. 1995. Is total mercury concentration a good predictor of methyl mercury concentration in aquatic systems. Water, Air Soil Pollut 80:715–724.

Kendall RJ, Brewer LW, Lacher TEJ, Whitten ML, Marden BT. 1989. The use of starling nest boxes for field reproductive studies. Institute of Wildlife Toxicology. EPA/600/ 8?89/056. National Technical Information Services publication number: PB89 195 028/AS, i–vii + 82 pp.

Kenow KP, Gutreuter S, Hines RK, Meyer MW, Fournier F, Karasov WH. 2003. Effects of methyl mercury exposure on the growth of juvenile common loons. Ecotoxicology 12:171–182.

Kerper LE, Ballatori N, Clarkson TW. 1992. Methylmercury transport across the blood-brain barrier by an amino acid carrier. Am J Physiol 262:R761–R765.

Kim CY, Nakai K, Kasanuma Y, Satoh H. 2000. Comparison of neurobehavioral changes in three inbred strains of mice prenatally exposed to methylmercury. Neurotoxicol Teratol 22:397–403.

Kim EY, Murakami T, Saeki K, Tatsukawa R. 1996. Mercury levels and its chemical form in tissues and organs of seabirds. Arch Environ Contamin Toxicol 30:259–266.

King KA, Custer TW, Weaver DA. 1994. Reproductive success of barn swallows nesting near a selenium-contaminated lake in east Texas, USA. Environ Pollut 84:53–58.

Kiparissis Y, Akhtar P, Hodson PV, Brown RS. 2003. Partiton-controlled delivery of toxicants: a novel *in vivo* approach for embryo toxicity testing. Environ Sci Technol 37:2262–2266.

Kishimoto T, Oguri T, Ueda D, Tada M. 1996. Methylmercury modulation of monocyte chemotactic protein-1 mRNA expression in human peripheral blood mononuclear cells. Hum Cell 9:371–374.

Knudsen TB, O'Hara MF, Charlap JH, Donahue RJ, Craig RC. 2000. Mitochondrial mechanisms of developmental toxicity: critical events and intervention. Toxicologist 54:69–70.

Koivusaari J, Nuuja I, Palokangas R, Finnlund M. 1980. Relationships between productivity, eggshell thickness and pollutant contents of addled eggs in the population of white-tailed eagles *Haliaetus albicilla L.* in Finland during 1969–1978. Environ Pollut 23:41–52.

Komsta-Szumska E, Czuba M, Reuhl KR, Miller DR. 1983. Demethylation and excretion of methyl mercury by the guinea pig. Environ Res 32:247–257.

Krabbenhoft DP, Gilmour CC, Benoit JM, Babiarz CL, Andren AW, Hurley JP. 1998. Methyl mercury dynamics in littoral sediments of a temperate seepage lake. Can J Fish Aquat Sci 55:835–844.

Kraus ML. 1989. Bioaccumulation of heavy metals in pre-fledgling tree swallows (*Tachycineta bicolor*). Bull Environ Contam Toxicol 43:407–414.

Kreps SE, Banzet N, Christiani DC, Polla BS. 1997. Molecular biomarkers of early responses to environmental stressors — implications for risk assessment and public health. Rev Environ Health 12:261–280.

Kucera E. 1983. Mink and otter as indicators of mercury in Manitoba waters. Can J Zool 61:2250–2256.

Law RJ. 1996. Metals in marine mammals. In: Beyer WN, Heinz GH, Redmon-Norwood AW, editors, Environmental contaminants in wildlife: interpreting tissue concentrations. Boca Raton (FL): CRC Press, p. 357–376.

Lee S, Yang RS. 2002. Interactive effects of methylmercury and PCB mixtures on neurodevelopment in mice. Toxicologist 66:132–133.

Leonzio C, Massi A. 1989. Metal biomonitoring in bird eggs. A critical experiment. Bull Environ Contam Toxicol 43:402–406.

Leonzio C, Monaci F. 1996. Multiresponse biomarker evaluation of interactions between methylmercury and Aroclor 1260 in quail. Ecotoxicology 5:365–376.

Lewandowski TA, Bartell SM, Ponce RA, Faustman EM. 2001. Mechanism-based comparison of *in vitro* and *in vivo* data on methylmercury toxicity in developing rodents. Toxicologist 60:258.

Lewis SA, Becker PH, Furness RW. 1993. Mercury levels in eggs, tissues and feathers of herring gulls (*Larus argentatus*) from the German Wadden Sea Coast. Environ Pollut 80:293–299.

Li S, Thompson SA, Woods JS. 1996. Localization of gamma-glutamylcysteine synthetase mRNA expression in mouse brain following methylmercury treatment using reverse transcription in situ PCR amplification. Toxicol Appl Pharmacol 140:180–187.

Lind B, Friberg L, Nylander M. 1988. Preliminary studies on methylmercury biotransformation and clearance in the brain of primates. II. Demethylation of mercury in brain. J Trace Elem Exp Med 1:49–56.

Loerzel SM, Samuelson DA, Sepulveda MS, Spalding MG, Lewis P, Smith PJ. 1996. Initial evaluation of mercury toxicity in the eye of the great egret. Investig Ophthalmol Vis Sci 37:s349.

Loerzel S, Samuelson D, Szabo N. 1999. Ocular effects methylmercury toxicity in juvenile double-crested cormonants (*Phalacrocorax auritus*). Investigative Ophthalmology & Visual Science. Annual meeting of the Association for Research in Vision and Ophthalmology. Fort Lauderdale, FL, May 9–14, 1999.

Loftus WF, Trexler JC, Jones RD. 1998. Mercury transfer through an Everglades aquatic food web. Dept. of Biol. Sci. and SE Environ. Res. Prog., Florida International University Final Report to the Florida Department of Environmental Protection, December 1998, contract SP-329.

Lord CG, Gaines KF, Boring CS, Brisbin ILJ, Gochfeld M, Burger J. 2002. Raccoon (*Procyon lotor*) as a bioindicator of mercury contamination at the U.S. Department of Energy's Savannah River Site. Archiv Environ Contam Toxicol 43:356–363.

Louis VL, Barlow JC. 1993. The reproductive success of tree swallows nesting near experimentally acidified lakes in northwestern Ontario. Can J Zool 71:1090–1097.

Maier WE, Costa LG. 1990. Sodium, potassium-ATPase in rat brain and erythrocytes as a possible target and marker, respectively, for neurotoxicity: studies with chlordecone, organotins and mercury compounds. Toxicol Lett 51:175–188.

Mallory ML. 1994. Notes on egg laying and incubation in the common merganser. Wilson Bull 106:757–759.

Martin AR, Reeves RR. 2002. Diversity and zoogeography. In: Hoelzel, AR, editor, Marine mammal biology: an evolutionary approach. Oxford, UK: Blackwell Science Ltd.

Mason CF, Ekins G, Ratford JR. 1997. PCB congeners, DDE, dieldrin and mercury in eggs from an expanding colony of cormorants (*Phalacrocorax carbo*). Chemosphere 34:1845–1849.

Mason RP, Laporte J, Andres S. 2000. Factors controlling the bioaccumulation of mercury, methylmercury, arsenic, selenium, and cadmium by freshwater invertebrates and fish. Arch Environ Contam Toxicol 38:283–297.

Mayne GJ, Bishop CA, Martin PA, Boermans HJ, Hunter B. 2005. Thyroid function in nestling tree swallows and eastern bluebirds exposed to non-persistent pesticides and p, p'-DDE in apple orchards of southern Ontario, Canada. Ecotoxicology 14:381–396.

McCrary JE, Heagler MG. 1997. The use of a simultaneous multiple species acute toxicity test to compare the relative sensitivities of aquatic organisms to mercury. J Environ Sci Health Part A: Environ Sci Eng Toxic Hazardous Substance Control 32:73–81.

McMurtry MJ, Wales DL, Scheider WA, Beggs GL, Dimond PE. 1989. Relationship of mercury concentrations in lake trout (*Salvelinus namaycush*) and smallmouth bass (*Micropterus dolomieui*) to the physical and chemical characteristics of Ontario lakes. Can J Fish Aquat Sci 46:426–434.

Medinsky MA, Klaassen CD. 1996. Toxicokinetics. In: Klaassen CD, editor, Casarett and Doull's toxicology: the basic science of poisons. New York: McGraw-Hill.

Meinelt T, Krueger R, Pietrock M, Osten R, Steinberg C. 1997. Mercury pollution and macrophage centres in pike (*Esox lucius*) tissues. Environ Sci Pollut Res Int 4:32–36.

Meyer MW. 1998. Ecological risk of mercury in the environment: the inadequacy of "the best available science." Environ Toxicol Chem 17:137–138.

Meyer MW, Evers DC, Daulton T, Braselton WE. 1995. Common loons (*Gavia immer*) nesting on low pH lakes in northern Wisconsin have elevated blood mercury content. Water Air Soil Pollut 80:871–880.

Meyer MW, Evers DC, Hartigan JJ, Rasmussen PS. 1998. Patterns of Common Loon (*Gavia immer*) mercury exposure, reproduction, and survival in Wisconsin, USA. Environ Toxicol Chem 17:184–191.

Meyers-Schoene L, Walton BT. 1990. Comparison of two freshwater turtle species as monitors of environmental contamination. Govt Reports Announcements & Index 24.

Meyers-Schone L, Shugart LR, Beauchamp JJ, Walton BT. 1993. Comparison of two freshwater turtle species as monitors of radionuclide and chemical contamination: DNA damage and residue analysis. Environ Toxicol Chem 12:1487–1496.

Mierle G, Addison EM, MacDonald KS, Joachim DG. 2000. Mercury levels in tissues of otters from Ontario, Canada: variation with age, sex, and location. Environ Toxicol Chem 19:3044–3051.

Mineau P, Fox GA, Norstrom RJ, Weseloh DV, Hallet DJ, Ellenton JA. 1984. Using the herring gull to monitor levels and effects of organochlorine contamination in the Canadian Great. Adv Environ Sci Technol 14:425–452.

Miura K, Koide N, Himeno S, Nakagawa I, Imura N. 1999. The involvement of microtubular disruption in methylmercury-induced apoptosis in neuronal and nonneuronal cell lines. Toxicol Appl Pharmacol 160:279–288.

Miura K, Koyama T, Nakamura I. 1978. Mercury content in museum and recent specimens of Chiroptera in Japan. Bull Environ Contam Toxicol 20:696–701.

Monteiro LR, Furness RW. 1995. Seabirds as monitors of mercury in the marine environment. Water Air Soil Pollut 80:851–870.

Monteiro LR, Furness RW. 2001. Kinetics, dose-response, and excretion of methylmercury in free-living adult Cory's shearwaters. Environ Sci Technol 35:739–746.

Monteiro LR, Granadeiro JP, Furness RW. 1998. Relationship between mercury levels and diet in Azores seabirds. Mar Ecol Progr Ser 166:259–265.

Moore DR, Teed RS, Richardson GM. 2003. Derivation of an ambient water quality criterion for mercury: taking account of site-specific conditions. Environ Toxicol Chem 22:3069–3080.

Moore DRJ, Sample BE, Suter GW, Parkhurst BR, Teed RS. 1999. A probabilistic risk assessment of the effects of methylmercury and PCBs on mink and kingfishers along East Fork Poplar Creek, Oak Ridge, Tennessee, USA. Environ Toxicol Chem 18:2941–2953.

Moreira CM, Oliveira EM, Bonan CD, Sarkis JJ, Vassallo DV. 2003. Effects of mercury on myosin ATPase in the ventricular myocardium of the rat. Comp Biochem Physiol C Toxicol Pharmacol 135C:269–275.

National Research Council. 1989. Biologic Markers of Air-Pollution Stress and Damage in Forests. National Academies Press: Committee on Biologic Markers of Air-Pollution Damage in Trees.

Nemeth KR, Charlap JH, O'Hara MF, Craig RC, Knudsen TB. 2002. Microarray analysis of developmental toxicity: ontogenetic profiles of susceptibility in the mouse embryonic eye. Toxicologist 66:28.

Newland MC, Paletz EM. 2000. Animal studies of methylmercury and PCBs: What do they tell us about expected effects in humans? Neurotoxicology Little Rock [print] December 21:1003–1028.

Newman J, Zillioux E, Rich E, Liang L, Newman C. 2004. Historical and other patterns of monomethyl and inorganic mercury in the Florida panther (*Puma concolor coryi*). Arch Environ Contam Toxicol 48:75–80.

Nichols J, Bradbury S, Swartout J. 1999. Derivation of wildlife values for mercury. J Toxicol Environ Health Part B 2:325–355.

Nichols JW, Larsen CP, McDonald ME, Niemi GJ, Ankley GT. 1990. Bioenergetics-based model for accumulation of polychlorinated biphenyls by nestling tree swallows, *Tachycineta bicolor*. Environ Sci Technol 29:604–612.

Nicholson JK, Osborn D. 1983. Kidney lesions in pelagic seabirds with high tissue levels of cadmium and mercury. J Zool London 200:99–118.

Nicholson JK, Osborn D. 1984. Kidney lesions in juvenile starlings *Sturnus vulgaris* fed on a mercury-contaminated synthetic diet. Environ Pollut Ser A 33:195–206.

Nielsen JB, Andersen O, Grandjean P. 1994. Evaluation of mercury in hair, blood and muscle as biomarkers for methylmercury exposure in male and female mice. Arch Toxicol 68:317–321.

Niimi AJ, Lowe-Jinde L. 1984. Differential blood cell ratios of rainbow trout (*Salmo gairdneri*) exposed to methylmercury and chlorobenzenes. Arch Environ Contam Toxicol 13:303–311.

Nisbet ICT. 1998. Trends in concentrations and effects of persistent toxic contaminants in the Great Lakes: their significance for inferring management actions. Environ Manage Assess 53:3–15.

Nisbet I, Montoya J, Burger J, Hatch J. 2002. Use of stable isotopes to investigate individual differences in diets and mercury exposures among common terns *Sterna hirundo* in breeding and wintering grounds. Mar Ecol Progr Ser 242:267–274.

Nocera JJ, Taylor PD. 1998. *In situ* behavioral response of common loons associated with elevated mercury (Hg) exposure. Conserv Ecol 2:10–17.

Nordenhall K, Dock L, Vahter M. 1998. Cross-fostering study of methyl mercury retention, demethylation and excretion in the neonatal hamster. Pharmacol Toxicol 82:132–136.

Norseth T, Clarkson TW. 1970. Studies on the biotransformation of Hg-labeled methylmercuric chloride in rats. Arch Environ Health 21:717–727.

O'Connor DJ, Nielsen SW. 1981. Environmental survey of methylmercury levels in wild mink (*Mustela vison*) and otter (*Lutra canadensis*) from the northeastern United States and experimental pathology of methylmercurialism in the otter. In: Chapman JD, Pursley D, editors, World Furbearer Conf Proc. Frostburg, MD, p. 1728–1745.

O'Hara TM, Woshner V, Bratton G. 2003. Inorganic pollutants in Arctic marine mammals. In: Vos JG, Bossart GD, Fournier M, O'Shea TJ, editors, Toxicology of marine mammals. London (UK): Taylor & Francis, p. 206–246.

O'Shea TJ. 1999. Environmental contaminants and marine mammals. In: Reynolds JEI, SA Rommel, editors, Biology of marine mammals, Washington, DC: Smithsonian Institution Press, p. 485–563.

O'Shea TJ, Tanabe S. 2003. Persistent ocean contaminants and marine mammals: a retrospective overview. In: Vos JG, Bossart GD, Fournier M, O'Shea TJ, editors, Toxicology of marine mammals, London (UK): Taylor & Francis, p. 99–134.

Odsjo T, Roos A, Johnels AG. 2004. The tail feathers of osprey nestlings (*Pandion haliaetus* L.) as indicators of change in mercury load in the environment of southern Sweden (1969–1998): a case study with a note on the simultaneous intake of selenium. Ambio 33:133–137.

Olsen AR, Sedransk J, Edwards D, Gotway CA, Liggett Rathbun S, Reckhow KH, Young LJ. 1999. Statistical issues for monitoring ecological and natural resources in the United States. Environ Monit Assess 54:1–45.

Olsen B, Evers DC, DeSorbo C. 2000. The effect of methylated mercury on the diving frequency of the common loon. J Ecol Res 2:67–72.

Omata S, Kasama H, Hasegawa H, Hasegawa K, Ozaki K, Sugano H. 1986. Species difference between rat and hamster in tissue accumulation of mercury after administration of methylmercury. Arch Toxicol 59:249–254.

Omata S, Toribara TY, Cernichiari E, Clarkson TW. 1988. Biotransformation of methylmercury *in-vitro* in the tissues of wild and laboratory animals. Fifty-ninth Annual Meeting of the Zoological Society of Japan, Sapporo, Japan, Zool Soc Tokyo, Japan.

Ou YC, White CC, Krejsa CM, Ponce RA, Kavanagh TJ, Faustman EM. 1999. The role of intracellular glutathione in methylmercury-induced toxicity in embryonic neuronal cells. Neurotoxicology Little Rock Oct 20:793–804.

Oxynos K, Schmitzer J, Kettrup A. 1993. Herring gull eggs as bioindicators for chlorinated hydrocarbons contribution to the German Federal Environmental Specimen Bank. Sci Total Environ 140:387–398.

Pekarik C, Weseloh DV. 1998. Organochlorine contaminants in herring gull eggs from the Great Lakes, 1974–1995: change point regression analysis and short-term regression. Environ Monit Assess 53:77–115.

Pekarik C, Weselo DV, Barret GC, Simon M, Bishop CA, Pettit KE. 1998. An atlas of contaminants in the eggs of fish-eating colonial birds of the Great Lakes (1993–1997), Vol 1 and 2. Canadian Wildlife Service, Ontario Region Tech. Rep. Series Number 322.

Pennuto CM, Lane OP, Evers DC, Taylor RJ, Loukmas J. 2005. Mercury in the northern crayfish *Orconectes virilis* (Hagen), in New England, USA. Ecotoxicology 14:149–162.

Pierotti RJ, Good TP. 1994. Herring Gull (*Larus argentatus*). In: Poole A, Gill F, editors, The birds of North America. No. 124.

Pingree SD, Simmonds PL, Woods JS. 2001. Effects of 2,3-dimercapto-1-propanesulfonic acid (DMPS) on tissue and urine mercury levels following prolonged methylmercury exposure in rats. Toxicol Sci 61:224–233.

Piper WH, Paruk JD, Evers DC, Meyer MW, Tischler KB, Klich M, Hartigan JJ. 1997. Local movements of color-marked common loons. J Wildl Manage 61:1253–1261.

Piper WH, Tischler KB, Klich M. 2000. Territory acquisition in loons: the importance of take-over. Anim Behav 59:385–394.

Ponce RA, Bartell SM, Kavanagh TJ, Woods JS, Griffith WC, Lee RC, Takaro TK, Faustman EM. 1998. Uncertainty analysis methods for comparing predictive models and biomarkers: a case study of dietary methyl mercury exposure. Regulatory Toxicol Pharmacol 28:96–105.

Poole AF. 1989. Osprey: a natural and unnatural history. New York: Cambridge University Press.

Porcella DB, Zillioux EJ, Grieb TM, Newman JR, West GB. 2004. Retrospective study of mercury in raccoons (*Procyon lotor*) in South Florida. Ecotoxicology 13:207–221.

Quinney TE, Smith PC. 1978. Reproductive success, growth of nestlings and foraging behaviour of the great blue heron (*Ardea herodias herodias* L.). Canadian Wildlife Service, Final contract report KL229-5-7077.

Rattner BA, McGowan PC, Golden NH, Hatfield JS, Toschik PC, Lukei Jr RF, Hale RC, Schmitz-Afonso I, Rice CP. 2004. Contaminant exposure and reproductive success of ospreys (*Pandion haliaetus*) nesting in Chesapeake Bay regions of concern. Arch Environ Contam Toxicol 47:126–140.

Rimmer C, McFarland K, Evers DC, Taylor T. 2005. Mercury levels in Bicknell's thrush and other insectivorous passerine birds in montane forests of northeastern United States. Ecotoxicology 14:223–240.

Risebrough RW. 1986. Pesticides and bird populations. Curr Ornithol 3:397–427.

Roberts TM, Hutchinson TC, Paciga J, Chattopadhyay A, Jervis RE, VanLoon J, Parkinson DK. 1974. Lead contamination around secondary smelters: estimation of dispersal and accumulation by humans. Science 186:1120–1123.

Roelke ME, Schultz DP, Facemire CF, Sundlof SF, Royals HE. 1991. Mercury contamination in Florida panthers. Eustis, FL: Florida Panther Interagency Committee, p. 54.

Rogers DW, Beamish FW. 1981. Uptake of waterborne methylmercury by rainbow trout (*Salmo gairdneri*) in relation to oxygen consumption and methylmercury concentration. Can J Fish Aquat Sci 38:1309–1315.

Ronald K, Tessaro SV, Uthe JF, Freeman HC, Frank R. 1977. Methylmercury poisoning in the harp seal (*Pagophilus groenlandicus*). Sci Tot Environ 8:1–11.

Rood BE, Gottgens JF, Delfino JJ, Earle CD, Crisman TL. 1995. Mercury accumulation trends in Florida Everglades and savannas marsh flooded soils. Water Air Soil Pollut 80:981–990.

Ross K, Abraham K, Gadawski T, Rempel R, Gabor T, Maher R. 2002. Abundance and distribution of breeding waterfowl in the Great Clay Belt of Northern Ontario. Can Field-Naturalist 116:42–50.

Rossi AD, Viviani B, Zhivotovsky B, Manzo L, Orrenius S, Vahter M, Nicotera P. 1997. Inorganic mercury modifies Ca^{2+} signal, triggers apoptosis and potentiates NMDA toxicity in cerebellar granule neurons. Cell Death Differentiation 4:317–324.

Rowe CL, Kinney OM, Nagle RD, Congdon JD. 1998. Elevated maintenance costs in an anuran (*Rana catesbeiana*) exposed to a mixture of trace elements during the embryonic and early larval periods. Physiolog Zool 71:27–35.

Ruckel S. 1993. Mercury concentrations in alligator meat in Georgia. Annual Conference of Southeast Association of Fish and Wildlife Agencies. 47:287–292.

Rudneva-Titova I. 1998. The biochemical effects of toxicants in developing eggs and larvae of Black Sea fish species. Mar Environ Res 46:499–500.

Ruelle R. 1992. Contaminant evaluation of interior least tern and piping plover eggs and chicks on the Missouri River, South Dakota. GRA & I Issue 4 residue NTIS/PB92-106210.

Rumbold DG, Niemczyk SL, Fink LE, Chandrasekhar T, Harkanson B, Laine KA. 2001. Mercury in eggs and feathers of great egrets (*Ardea albus*) from the Florida Everglades. Arch Environ Contam Toxicol 41:501–507.

Russell D. 2003. Evaluation of the Clean Water Act Section 304(a) human health criterion for methylmercury; protectiveness for threatened and endangered wildlife in California. Department of the Interior, US Fish and Wildlife Service DW-14-95556801-0.

Scheuhammer AM. 1987. The chronic toxicity of aluminum, cadmium, mercury, and lead in birds: a review. Environ Pollut 46:263–295.

Scheuhammer AM. 1988. Chronic dietary toxicity of methylmercury in the zebra finch, *Poephila guttata*. Bull Environ Contam Toxicol 40:123–130.

Scheuhammer AM. 1990. Accumulation and toxicity of mercury, cadmium and lead in vertebrates. In: Workshop to Design Baseline and Monitoring Studies for the OCS Mining Program, Norton Sound, Alaska — Workshop Proceedings, US Dept. of the Interior: Minerals Management Service, OCS Study, mms90-059.

Scheuhammer AM, Atchison CM, Wong AHK, Evers DC. 1998a. Mercury exposure in breeding common loons (*Gavia immer*) in central Ontario, Canada. Environ Toxicol Chem 17:191–196.

Scheuhammer AM, Wong AH, Bond D. 1998b. Mercury and selenium accumulation in common loons (*Gavia immer*) and common mergansers (*Mergus merganser*) from eastern Canada. Environ Toxicol Chem 17:197–201.

Scheuhammer AM, Perrault JA, Bond DE. 2001. Mercury, methylmercury, and selenium concentrations in eggs of Common Loons (*Gavia immer*) from Canada. Environ Monitor Assess 72:79–74.

Schreiber BA, Burger J. 2002. Biology of marine birds. Boca Raton (FL):CRC Press.

Schwarzbach SE, Henderson J, Albertson J, Hofius J. 1997. Assessing risk from methylmercury in tidal marsh sediments to reproduction of California Clapper Rails (*Rallus longirostris obsoletus*) using target diet and target egg concentration approaches. Poster for Society of Environmental Toxicology and Chemistry, November 1997, San Francisco, CA.

Seegal RF, Bemis JC. 2000. Polychlorinated biphenyls and methylmercury synergistically alter dopamine and intracellular calcium. Neurotoxicology 21:243.

Seigneur C, Vijayaraghavan K, Lohman K, Karamchandani P, Scott C. 2004. Global source attribution for mercury deposition in the United States. Environ Sci Technol 38:555–569.

Sepulveda MS, Poppenga RH, Arrecis JJ, Quinn LB. 1998. Mercury and selenium concentrations in free-ranging in tissues from double-crested cormorants (Phalacrocorax auritus) from southern Florida. Colonial Waterbirds 21:35–42.

Shanker G, Mutkus LA, Walker SJ, Aschner M. 2002. Methylmercury enhances arachidonic acid release and cytosolic phospholipase A2 expression in primary cultures of neonatal astrocytes. Brain Res Mol Brain Res 106:1–11.

Shaw BP, Dash S, Panigrahi AK. 1991. Effect of methyl mercuric chloride treatment on haematological characteristics and erythrocyte morphology of Swiss mice. Environ Pollut 73:43–52.

Shenker BJ, Guo TL, Shapiro IM. 1998. Low-level methylmercury exposure causes human T-cells to undergo apoptosis: evidence of mitochondrial dysfunction. Environ Res 77:149–159.

Shipp AM, Gentry PR, Lawrence G, Van Landingham CF, Covington T, Clewell HJ, Gribben K, Crump K. 2000. Determination of a site-specific reference dose for methylmercury for fish-eating populations. Toxicol Ind Health 16:335–438.

Sillman AJ, Weidner WJ. 1993. Low levels of inorganic mercury damage the corneal endothelium. Exper Eye Res 57:549–555.

Slotton DG, Reuter JE, Goldman CR. 1995. Mercury uptake patterns of biota in a seasonally anoxic northern California Reservoir. Water, Air Soil Pollut 80:841–850.

Spalding MG, Frederick PC, McGill HC, Bouton SN, McDowell LR. 2000a. Methylmercury accumulation in tissues and its effects on growth and appetite in captive great egrets. J Wildl Dis 36:411–422.

Spalding MG, Frederick PC, McGill HC, Bouton SN, Richey LJ, Schumacher IM, Blackmore CG, Harrison J. 2000b. Histologic, neurologic, and immunologic effects of methylmercury in captive great egrets. J Wildl Dis 36:423–435.

Stafford CP, Haines TA. 2001. Mercury contamination and growth rate in two piscivore populations. Environ Toxicol Chem 20:2099–2101.

Stalmaster MV. 1987. The bald eagle. New York (NY): Universe Books.

Steidl RJ, Griffin CR, Niles LJ. 1991. Contaminant levels of osprey eggs and prey reflect regional differences in reproductive success. J Wildl Manage 55:601–608.

Stendell RC, Ohlendorf HM, Klaas EE, Elder JB. 1976. Mercury in eggs of aquatic birds, Lake St. Clair — 1973. Pestic Monit J 10:7–9.

Stewart FM, Phillips RA, Bartle JA, Craig J, Shooter D. 1999. Influence of phylogeny, diet, moult schedule and sex on heavy metal concentrations in New Zealand Procellarii-formes. Mar Ecol Progr Ser 178:295–305.

Struger J, Elliott JE, Weseloh DV. 1987. Metals and essential elements in herring gulls from the Great Lakes 1983, USA Canada. J Great Lakes Res 13:43–55.

Sundlof SF, Spaulding MG, Wentworth JD, Steible CK. 1994. Mercury in livers of wading birds (Ciconiiformes) in Southern Florida. Arch Environ Contam Toxicol 27:299–305.

Takeuchi T. 1977. Neuropathology of Minimata disease in Kumamoto: especially at the chronic stage. In: Roizin L, Shiraki H, Grcevic N, editors, Neurotoxicology. New York: Raven Press.

Tansy CL. 2002. A comparison of contaminants in tissues of mink (*Mustela vision*) from South Carolina and Louisiana. MS thesis, Clemson University.

Tatara CP, Mulvey M, Newman MC. 2002. Genetic and demographic responses of mercury-exposed mosquitofish (*Gambusia holbrooki*) populations: temporal stability and reproductive components of fitness. Environ Toxicol Chem 21:2191–2197.

Teigen SW, Andersen RA, Daae HL, Skaare JU. 1999. Heavy metal content in liver and kidneys of grey seals (*Halichoerus grypus*) in various life stages correlated with metallothionein levels: some metal-binding characteristics of this protein. Environ Toxicol Chem 18:2364–2369.

Thaxton JP, Parkhurst CR. 1973. Abnormal mating behavior and reproductive dysfunction caused by mercury in Japanese quail. Proc Soc Exper Biol Med 144:252–255.

Thompson DR. 1996. Mercury in birds and terrestrial mammals. In: Beyer WN, Heinz GH, Redmon-Norwood AW, editors, Environmental contaminants in wildlife: interpreting tissue concentrations. Boca Raton (FL): Lewis Publishers.

Thompson DR, Becker PH, Furness RW. 1993. Long-term changes in mercury concentrations in herring gulls *Larus argentatus* and common terns *Sterna hirundo* from the German North Sea coast. J Appl Ecol 30:316–320.

Thompson DR, Furness RW. 1989a. The chemical form of mercury stored in South Atlantic seabirds. Environ Pollut 60:305–318.

Thompson DR, Furness RW. 1989b. Comparison of the levels of total and organic mercury in seabird feathers. Mar Pollut Bull 20:557–579.

Thompson DR, Furness RW, Walsh PM. 1992. Historical changes in mercury concentrations in the marine ecosystem of the north and northeast Atlantic Ocean as indicated by seabird feathers. J Appl Ecol 29:79–84.

Thompson DR, Hamer KC, Furness RW. 1991. Mercury accumulation in great skuas, *Cathar-acta skua* of known age and sex and its effects upon breeding and survival. J Appl Ecol 28:672–684.

Timken RL, Anderson BW. 1969. Food habits of common mergansers in the northcentral United States. J Wildl Manage 33:87–91.

Toschik PC, Rattner BA, McGowan PC, Christman MC, Carter DB, Hale RC, Matson CW, Ottinger MA. 2005. Effects of contaminant exposure on reproductive success of ospreys (*Pandion haliaetus*) nesting in Delaware River and Bay, USA. Environ Toxicol Chem 24:617–628.

Tsuchiya W, Okada Y. 1982. Differential effects of cadmium and mercury on amino acid and sugar transport in the bullfrog small intestine. Experientia (Basel) 38:1073–1075.

[USCOE] US Army Corps of Engineers and [USEPA] US Environmental Protection Agency. 2005. Environmental Residue-Effects Database (ERED).

[USEPA] US Environmental Protection Agency. 1993a. A protocol for evaluating historical contamination of mercury in the Florida panther and other Florida wildlife. USEPA, Ecological Effects Branch KBN Engineering, Gainesville, FL; Final report 13341B1.

[USEPA] US Environmental Protection Agency. 1993b. Wildlife exposure factors handbook, Vol. 1. United States Environmental Protection Agency, Method EPA 600-R-93-187a.

[USEPA] US Environmental Protection Agency. 1995. Final water quality guidance for the great lakes system. Fed Reg 40 CFR Parts 9, 122, 123, 131, & 132 60:15366–15424.

[USEPA] US Environmental Protection Agency. 1997. Mercury study report to Congress, Vol VI: An ecological assessment of anthropogenic mercury emissions in the United States. Office of Air Quality Planning and Standards and Office of Research and Development, U.S. Environmental Protection Agency, EPA-452/R-97-008.

[USEPA] US Environmental Protection Agency. 2001. Water quality criterion for the protection of human health: methylmercury. EPA-823-R-01-001.

Van Der Molen EJ, Blok AA, Graaf GJ. 1982. Winter starvation and mercury intoxication in Gray Herons *(Ardea cinerea)* in the Netherlands. Ardea 70:173–184.

Vos P, Meelis E, Ter Keurs WJ. 2000. A framework for the design of ecological monitoring programs as a tool for environmental and nature management. Environ Monit Assess 61:317–344.

Wagemann RE, Trebacz G, Boila G, Lockhhart WL. 2000. Mercury species in the liver of ringed seals. Sci Total Environ 261:21–32.

Wagemann R, Innes S, Richard PR. 1996. Overview and regional and temporal differences of heavy metals in Arctic whales and ringed seals in the Canadian Arctic. Sci Total Environ 186:41–66.

Wagemann R, Trebacz E, Boila G, Lockhart WL. 1998. Methylmercury and total mercury in tissues of arctic marine mammals. Sci Total Environ 218:19–31.

Wallin K. 1984. Decrease and recovery patterns of some raptors in relation to the introduction and ban of alkyl-mercury and DDT in Sweden. Ambio 13:263–265.

Walsh PM. 1990. The use of seabirds as monitors of heavy metals in the marine environment. In: Furness RW, Rainbow PS, editors, Heavy metals in the marine environment. Boca Raton (FL): CRC Press.

Wayland M, Gilchrist HG, Marchant T, Keating J, Smits JE. 2002. Immune function, stress response, and body condition in Arctic-breeding common eiders in relation to cadmium, mercury, and selenium concentrations. Environ Res 90:47–60.

Wayland M, Smits JE, Gilchrist HG, Marchant T, Keating J. 2003. Biomarker responses in nesting, common eiders in the Canadian arctic in relation to tissue cadmium, mercury and selenium concentrations. Ecotoxicology 12:225–237.

Webber HM, Haines TA. 2003. Mercury effects on predator avoidance behavior of a forage fish, golden shiner *(Notemigonus crysoleucas)*. Environ Toxicol Chem 22:1556–1561.

Weech SA, Wilson LK, Langelier KM, Elliott JE. 2003. Mercury residues in livers of bald eagles *(Haliaeetus leucocephalus)* found dead or dying in British Columbia, Canada (1987–1994). Arch Environ Contam Toxicol 45:562–569.

Weiss-Penzias P, Jaffe DA, McClintick A, Prestbo EM, Landis MS. 2003. Gaseous elemental mercury in the marine boundary layer: evidence for rapid removal in anthropogenic pollution. Environ Sci Technol 37:3755–3763.

Welch LJ. 1994. Contaminant burdens and reproductive rates of bald eagles breeding in Maine. Master's Thesis: University of Maine. Orono, ME.

Wiemeyer SN, Bunck CM, Stafford CJ. 1993. Environmental contaminants in bald eagle eggs: 1980–1984 and further interpretations of relationships to productivity and shell thickness. Arch Environ Contam Toxicol 24:213–227.

Wiemeyer SN, Lamont TG, Bunck CM, Sindelar CR, Gramlich FJ, Fraser JD, Byrd MA. 1984. Organochloride pesticide, polychlorobiphenyl, and mercury residues in bald eagle eggs — 1969–1979 — and their relationships to shell thinnings and reproduction. Arch Environ Contam Toxicol 13:529–549.

Wiemeyer SN, Schmeling SK, Anderson A. 1987. Environmental pollutant and necropsy data for ospreys from the eastern United States, 1975–1982. J Wildl Dis 23:279–291.

Wobeser G. 1975. Prolonged oral administration of methyl mercury chloride to rainbow trout. J Fish Res Board Can 32:2015–2023.

Wobeser G, Nielsen NO, Schiefer B. 1976. Mercury and mink. I. Use of mercury-contaminated fish as a food for ranch mink intoxication. Can J Comp Med 40:30–33.

Wolfe MF, Kendall RJ. 1998. Age-dependent toxicity of terbufos and diazinon to European starlings (*Sturnis vulgaris*) and red-winged blackbirds (*Agelaius phoeniceus*). Environ Toxicol Chem 17:1300–1312.

Wolfe M, Norman D. 1998a. Effects of waterborne mercury on terrestrial wildlife at Clear Lake: Evaluation and testing of a predictive model. Environ Toxicol Chem 17:214–227.

Wolfe MF, Schwarzbach S, Sulaiman RA. 1998. The effects of mercury on wildlife: a comprehensive review. Environ Toxicol Chem 17:146–160.

Wood CM, Adams WS, Ankley GT, DiBona DR, Luoma SN, Playle RC, Stubblefield WC, Bergman HL, Erickson HL, Dorward-King EJ. 1997. Environmental toxicology of metals. In: Reassessment of metals criteria of aquatic life protection, Berman HL, Dorland-King DJ, editors. Pensacola (FL): SETAC, p. 31–51.

Wood PB. 1993. Mercury concentrations in blood and feathers of nestling Florida bald eagles. Annual Meeting of the Raptor Research Foundation, Inc, Charlotte, NC, November 28:68.

Wood PB, White JH, Steffer A, Wood JM, Facemire CF, Percival HF. 1996. Mercury concentrations in tissues of Florida bald eagles. J Wildl Manage 60:178–185.

Wren CD. 1984. Distribution of metals in tissues of beaver, raccoon and otter from Ontario, Canada. Sci Tot Environ 34:177–184.

Wren CD. 1986. A review of metal accumulation and toxicity in wild mammals. I. Mercury. Environ Res 40:1737–1744.

Wren CD, Hunter DB, Leatherland JF, Stokes PM. 1987a. The effects of polychlorinated biphenyls and methylmercury, singly and in combination on mink. 2. Reproduction and kit development. Arch Environ Contam Toxicol 16:449–454.

Wren CD, Hunter DB, Leatherland JF, Stokes PM. 1987b. The effects of polychlorinated biphenyls and methylmercury, singly and in combination, on mink. I. Uptake and toxic responses. Arch Environ Contam Toxicol 16:441–447.

Wren CD, Stokes PM, Fischer KL. 1986. Mercury levels in Ontario Canada mink and otter relative to food levels and environmental acidification. Can J Zool 64:2854–2859.

Yang MG, Krawford KS, Gareia JD, Wang JHC, Lei KY. 1972. Deposition of mercury in fetal and maternal brain. Proc Soc Exp Biol Med 141:1004–1007.

Yanochko GM, Jagoe CH, Brisbin Jr IL. 1997. Tissue mercury concentrations in alligators (*Alligator mississippiensis*) from the Florida Everglades and the Savannah River Site, South Carolina. Arch Environ Contam Toxicol 32:323–328.

Yasutake A, Hirayama K. 1994. Acute effects of methylmercury on hepatic and renal glutathione metabolisms in mice. Arch Toxicol 68:512–516.

Yasutake A, Nakano A, Miyamoto K, Eto K. 1997. Chronic effects of methylmercury in rats. I. Biochemical aspects. Tohoku J Exp Med 182:185–196.

Yates DE, Mayack DT, Munney K, Evers DC, Major A, Kaur T, Taylor RJ. 2005. Mercury levels in mink (*Mustela vison*) and river otter (*Lontra canadensis*) from northeastern North America. Ecotoxicology 14:263–274.

Yediler A, Jacobs J. 1995. Synergistic effects of temperature; oxygen and water flow on the accumulation and tissue distribution of mercury in carp (*Cyprinus carpio* L.). Chemosphere 31:4437–4453.

Yuska DE, Skelly JM, Ferdinand JA, Stevenson RE, Savage JE, Mulik JD, Hines A. 2003. Use of bioindicators and passive sampling devices to evaluate ambient ozone concentrations in north central Pennsylvania. Environ Pollut 125:71–80.

Zillioux EJ, Newman JR. 2003. Bioindicators — essential tools for realistic assessment and remediation cost control. Soil, Sediment and Water 9:11.

Zillioux EJ, Porcella DB, Benoit JM. 1993. Mercury cycling and effects in freshwater wetland ecosystems. Environ Toxicol Chem 12:2245–2264.

6 An Integrated Framework for Ecological Mercury Assessments

Tamara Saltman, Reed Harris, Michael W. Murray, and Rob Reash

6.1 INTRODUCTION

Environmental data capable of demonstrating changes and trends in contaminants of concern to scientists and policymakers are crucial to understand the benefits of past and future regulatory and voluntary actions. While there will always be uncertainty associated with observed trends and the potential benefits of remedial actions, high-quality environmental data will enhance management and policy development capabilities. In addition, defensible environmental trend data are needed to assess the benefits society receives or can expect to receive in return for the resources invested in pollution control measures. In short, without good environmental monitoring, we may never know if we are making sound stewardship decisions.

Mercury (Hg) contamination is widespread in water, in surficial soils and sediments, and in the tissues of plants and animals in ecosystems around the globe. Once deposited to terrestrial and aquatic ecosystems, some inorganic mercury is transformed into methylmercury (MeHg), a highly toxic compound that bioaccumulates efficiently in food webs (Wiener et al. 2003). As a result of the toxicity of MeHg to wildlife and humans, many nations are interested in reducing environmental mercury contamination and associated biotic exposure (UNEP 2002).

While mercury is a naturally occurring element that cannot be eliminated from the environment, human activities and processes have greatly increased its abundance in the atmosphere and in terrestrial and aquatic environments (Jackson 1997; USEPA 1997; Fitzgerald et al. 1998), primarily by transporting this mercury from geological strata to the surface of the Earth and atmosphere (e.g., due to mining and coal combustion). One approach for reducing exposure to mercury in the industrialized world is to reduce anthropogenic emissions of the metal into the atmosphere. This has been done in some instances. For example, in the United States, regulation of medical waste incinerators and municipal waste combustors and steps taken to reduce use of mercury in products, as well as some reductions from power plants as a co-benefit of SO_2 emissions control, have already reduced mercury emissions. As a result, decreases in the accumulation of total Hg in depositing sediments have been measured in some lakes (Engstrom and Swain 1997; Kamman and Engstrom 2002).

191

Trends in mercury concentration in freshwater fish and loons as a result of changes in mercury deposition have also been reported (Fevold et al. 2002; Hrabik and Watras 2001). However, as the biogeochemical cycling of Hg is complex, it is necessary to comprehensively examine temporal patterns in mercury deposition, contamination, and bioaccumulation over susceptible ecosystems in many parts of the country to determine if the goal of reducing exposure has been met at regional, national, or international scales. There is currently no monitoring network or group of monitoring networks that can answer these questions. The purpose of this book is to provide guidance toward the development of such a network.

The 4 preceding chapters identify the indicators considered to be most useful for measuring trends in mercury contamination in specific ecosystem compartments: airsheds and watersheds, aquatic ecosystems, aquatic biota, and wildlife. Each chapter discusses the appropriate use and measurement of the recommended indicators. The chapters also discuss the application of monitoring data for defensibly assessing whether observed changes in mercury concentrations are associated with changes in mercury emissions and deposition. Factors confounding the ability to make such links are included in the discussions. The purpose of this chapter is to integrate the information in the 4 preceding chapters into a single set of indicators within a monitoring framework capable of assessing national or continental trends in mercury contamination in relation to observed trends in anthropogenic emissions.

6.2 RECURRING THEMES

In each chapter, the question of how to measure changes and trends in mercury contamination in the environment is approached from a similar perspective. This chapter discusses 5 unifying themes that emerged from the preceding chapters.

The first theme is integration. Each group of chapter authors recognized the importance of integrating the information from indicators across multiple ecosystem compartments and sampling locations. Individual indicators may provide information about concentration changes in a specific ecosystem compartment, but an integrated program involving mercury measurements and ancillary supporting data is needed to critically evaluate possible relationships between mercury emissions, deposition, and concentrations in key ecosystem compartments.

The importance of setting a baseline as promptly as possible from which to assess future changes was also recognized as an important component of the monitoring framework. It is important to be able to compare environmental conditions before and after emission reductions begin to fully capture the ecosystem changes that may be attributable to the emission reductions. If ecosystem monitoring does not begin until after emission reductions occur, the full benefit of the emission reductions may not be quantified directly (although indirect estimates from cores or archived samples may still be possible). In addition, although there is uncertainty regarding the time lags before detection of measured responses in an ecosystem, there are indications that in at least some ecosystems, detectable changes begin relatively quickly (Chapter 3, Chapter 5, Atkeson et al. 2003). Due to the need to compare future monitoring results to conditions before emission reductions begin, the chapter authors agreed that existing data and monitoring frameworks should be utilized whenever and wherever feasible and appropriate.

Perhaps most importantly, there was strong agreement on the need for standard monitoring protocols. This was one of the driving forces behind this book. Without standard monitoring protocols, it will be difficult — or impossible — to compare data collected from different locations or at different times, a capability that is absolutely critical for determining temporal and spatial trends. The scale, location, timing, and frequency of sampling are all important considerations that can make or break a monitoring strategy. While this book does not provide detailed sampling methods, the preceding chapters have provided important general guidance on approaches to sampling. Adherence to a coordinated sampling framework, similar to that presented here, is critical to ensure that the information gathered is useful for addressing key policy assessment questions. While the authors recognize that standard monitoring protocols can be difficult to implement and may complicate an organization's or researcher's ability to develop ideal site-specific monitoring protocols, the benefits of standardization into a single integrated monitoring network should far outweigh their limitations.

The fourth theme is the importance of ancillary data to go with each mercury indicator. The preceding chapters discuss many of the factors that can affect mercury concentrations in ecosystem compartments, including atmospheric and aquatic chemistry, hydrology, climate, and trophic conditions. Mercury loading rates can also affect mercury concentrations. The effect of inorganic mercury loading on MeHg concentrations in biota has been documented in cases of extreme contamination (e.g., chlor-alkali facilities), but the relationship is more difficult to isolate for cases involving smaller changes in loading, such as variations in atmospheric deposition to remote areas (USEPA 1997). This is partly because of the confounding effects of other factors (Chapters 2 through 5, Schindler et al. 1995). Therefore, the scope of the proposed trend-monitoring program should include measurements of the necessary ancillary data that can provide linkages between mercury emissions, atmospheric deposition, and concentrations in biota.

The fifth and final theme is the importance of models. The chapter authors support the appropriate use of models and co-located research programs to complement monitoring efforts. Models can help interpolate monitoring results between data collection points; for example, statistical models can be used to create smooth coverages of atmospheric deposition rates from a set of monitoring points (e.g., Grimm and Lynch 1991). Models can also be used to predict future changes based on existing data to inform policy decisions or actions (USEPA 1997). Mechanistic models can be used to simultaneously consider a range of factors affecting mercury concentrations and help to isolate the effects of mercury loading rates from other variables. Data from intensive monitoring locations can be useful for model development, and modeling data needs should be considered when designing a monitoring strategy.

6.2.1 DESIGN OF THE MONITORING NETWORK

The indicators presented in this monitoring framework are meant to function as pieces of an integrated assessment program. An integrated program is essential for

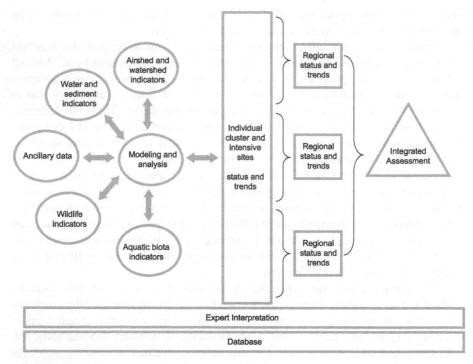

FIGURE 6.1 Integration of indicator framework.

defensible interpretation of linkages between atmospheric Hg deposition and MeHg concentrations in biota. While some of the indicators recommended here have stand-alone value as endpoints of concern, others are most useful when trying to understand the links and pathways between atmospheric Hg deposition and MeHg in biota. The interpretation of many of the indicators proposed here will rely in part on the collection of ancillary data, such as water chemistry (e.g., pH, dissolved organic carbon (DOC), and sulfate), as well as information on the mercury indicators themselves. In addition, atmospheric deposition monitoring must be coordinated with water and sediment sampling, and both of those must be coordinated with sampling of biota. Such integration is considered the foundation of the monitoring framework. Figure 6.1 is a conceptual framework of how these indicators can be integrated into a comprehensive assessment of ecosystem status and its response to changes in atmospheric mercury loadings as well as other influences.

It would be simpler to design monitoring programs if we knew *a priori* how quickly various environmental compartments are likely to respond to changes in mercury loadings. The indicators recommended in this book are based on current understanding of mercury cycling and transformation processes, which may be revised as more data become available, as well as the current state of methods for sampling and analysis. This long-term monitoring program should be designed to adapt in response to new information and to incorporate new methods without compromising the continuous long-term nature of the data collected. A discussion and synthesis of the criteria for selecting the indicators recommended in each chapter are presented below.

6.2.1.1 Criteria for Selection of Indicators

The recommended priorities of the monitoring program are to 1) establish a set of indicators that could be monitored to determine whether mercury concentrations in air, land, water, and biota are changing systematically with time, and 2) provide guidance regarding additional monitoring needed to help determine whether observed changes in mercury concentrations are related to reductions in mercury emissions. It is not the intent of this monitoring program to provide a complete understanding of mercury behavior in the environment or to measure mercury concentrations in every conceivable compartment in the environment.

To this point, this book has not addressed one of the primary barriers to comprehensive environmental monitoring: the limited funding available to implement and sustain such a program. Therefore, this chapter identifies the indicators for 2 levels of assessment that differ in the resources needed to implement and sustain them. One level focuses on documenting changes in mercury concentrations in the environment, while the second is expanded in scope to explore mechanistic linkages between changes in atmospheric mercury emissions, deposition rates, and soil, water, or biotic concentrations. *Both* programs have been developed with economic efficiencies as a consideration.

The first assessment program, termed the "trend only" approach, would determine trends in mercury contamination without focusing on causal factors. These indicators are here identified as "A" indicators. This approach would apply a core set of indicators to document mercury concentrations in key environmental compartments and would include minimal supporting ancillary data. These initial indicators simply answer the question "Has there been a change?" The second assessment program includes the A indicators but would be expanded in scope to more critically assess the role of emissions changes in observed changes in environmental mercury contamination. The indicators used to answer the question "What is causing the changes?" are identified here as "B" indicators. The purpose of this more scientifically rigorous question is to develop a suite of indicators that can more fully capture ecological changes and identify the probable cause for those changes. These indicators would significantly enhance our ability to defensibly interpret the trends identified with A indicators.

This framework represents a group consensus of the best way to transform, expand, and develop new and existing monitoring programs into an integrated assessment framework. The previous chapters in this book identified and discussed many attributes of a successful indicator (Chapters 2, 3, 4, and 5). The following attributes were identified:

- Relevance to public or ecological health
- Sensitivity to changes in mercury loading
- Availability of historical data
- Part of an existing data collection infrastructure

- Feasible methods for standard data collection
- Important link in the pathway between mercury emissions and methyl-mercury in biota
- Ability to isolate impacts of confounding factors other than changes in mercury emissions and deposition

For biological indicator species, several additional criteria are pertinent. The chosen indicators should:

- Have a broad geographic distribution, especially if comparison of ecosystem changes in different regions of the country is desired
- Have a well-known life history
- Show reasonable response time(s) to changes in mercury contamination
- Have resilience within the population to repetitive sampling, due either to the large number of individuals in the population or the availability of noninvasive sampling techniques

The indicators recommended in each chapter are compiled in Table 6.1. The criteria used by each set of chapter authors to rate their indicators have been summarized and the degree to which each indicator meets those criteria are identified. It should be noted that there is no perfect indicator, but the recommended indicators meet enough criteria that, in the opinion of the authors of these chapters, they are useful for assessment purposes. Detailed descriptions of the indicators and, in some cases, additional specific criteria are included in Chapters 2 through 5. The indicators are further subdivided according to their ability to measure environmental changes (A indicators) or to help assess the influence of atmospheric deposition (B indicators) on changes in other biotic and abiotic mercury concentrations in Tables 6.2 and 6.3, respectively.

6.2.2 CONSIDERATIONS FOR SAMPLING

As pointed out by numerous authors, detecting both spatial and temporal trends in environmental data is no trivial task, and numerous factors must be considered both in designing a robust sampling scheme and in interpreting data (e.g., Stow et al. 1998; Urquhart et al. 1998; Olsen et al. 1999; Larsen et al. 2001; Griffith et al. 2001; Kincaid et al. 2004). A central issue is having the statistical power to detect trends across time. The framework described by Urquhart et al. (1998) and Larsen et al. (2001) identified 4 types of variability in data that can influence the ability to detect spatial and temporal trends: 1) within year (i.e., short-term temporal, spatial, and measurement variation seen at an array of sites during a sampling window; 2) across-year concordant (variation seen at all sites in a region due to yearly changes in regional phenomena); 3) across-year independent (or interaction variation, capturing site-specific phenomena that can vary annually); and 4) among site (capturing fundamental differences between sites, due to, for example, landscape differences).

The primary considerations discussed in this book for sampling the indicators include the scale of measurements needed; the type of sampling location (e.g., undisturbed sites, clustered sites); the frequency of sampling (e.g., hourly, weekly, annually, biennially); and the duration of sampling needed to detect trends.

TABLE 6.1
Matrix of indicators identified as having the best potential for use in the context of measuring changes in environmental mercury contamination

Indicator	Criteria										Type
	1	2	3	4	5	6	7	8	9	10	
Chapter 2											
THg Wet deposition	*	*****	****	*****	*****	****	****	*****	NA	*****	B
THg in throughfall	*	*****	*	**	*****	****	****	***	NA	****	A
THg in litterfall; MeHg in litterfall and throughfall	*	***	*	***	*****	****	**	***	NA	*****	B
THg in snowpack	*	***	*	***	**	*****	****	*	NA	*****	B
Continuous speciated atmospheric Hg	*	*****	**	**	****	*****	****	*****	NA	****	B
Soil (THg and MeHg)	*	*	*	*	**	***	**	*****	NA	**	B
Soils solutions (THg and MeHg)	*	*	*	**	*	***	**	*****	NA	**	B
Forest floor surveys	*	*	*	**	**	***	**	***	NA	***	B
Chapter 3											
Streamwater (THg and MeHg)	***	***	**	****	*****	***	***	*****	NA	*****	A
THg in sediment	**	**	****	****	****	****	****	****	NA	*****	B
MeHg in sediment	****	***	**	****	****	***	***	****	NA	*****	B
% MeHg in sediments and soils	****	*	*	****	****	***	***	****	NA	*****	B
Instantaneous sediment methylation rate	**	****	*	*	*	**	**	**	NA	*****	B
THg accumulation in cores	***	**	****	***	***	****	****	*****	NA	*****	B
THg in surface water	**	***	***	****	****	****	***	****	NA	*****	B
MeHg in surface water (still water: lakes, reservoirs, wetlands)	****	***	***	****	****	***	***	****	NA	*****	B
Chapter 4											
MeHg in estuarine benthic invertebrates	*****	*	**	***	*****	*	***	*****	**	*****	B
MeHg and THg in prey fish	****	*****	*****	***	*****	****	*****	*****	*****	*****	A
Piscivorous fish	*****	***	*****	*****	*****	***	****	*****	***	****	A

TABLE 6.1 (continued)

Matrix of indicators identified as having the best potential for use in the context of measuring changes in environmental mercury contamination

Indicator	1	2	3	4	5	6	7	8	9	10	Type
Chapter 5											
(Terrestrial) Raccoon: THg in hair of adults	***	***	***	***	***	***	***	***	***	***	A, B
(Terrestrial) Bicknell's thrush: THg in blood of adults	**	*****	***	*****	*****	***	**	**	***	*****	A, B
(Riverine) Mink: THg in hair of adults	****	***	***	***	***	***	***	***	***	***	B
(Lake) Common loon: THg in blood of adults, also eggs	*****	***	*****	*****	*****	****	****	****	*****	*****	A, B
(Lake/coastal) Herring gull: THg in feathers and blood of adults; eggs	***	***	***	****	*****	***	***	**	*****	***	A, B
(Lake/coastal) Common tern: THg in feathers, blood of adults; eggs	****	***	****	****	*****	****	****	****	****	***	A, B
(Wetland) Tree swallow: THg in blood of adults	*	***	***	****	*****	****	***	*****	*****	*****	B
(Estuarine) Saltmarsh sharp-tailed sparrow and Seaside sparrow: THg in blood of adults	****	****	***	****	*****	****	***	**	*****	*****	B
(Marine nearshore) Harbor porpoise: THg in muscle of adult	***	**	****	****	****	**	**	***	****	**	A, B
(Marine offshore) Leach's storm-petrel: THg in blood of adult and juvenile	**	*****	***	***	*****	****	**	*****	***	*****	A, B
(Comparison across aquatic habitats) Belted kingfisher: THg in blood of adult and fledged young; eggs	****	***	****	*****	*****	****	***	*****	***	*****	B

Criteria: 1) Relevance to human health endpoints. 2) Sensitivity to change in loadings. 3) Overall historical data quality. 4) Data collection infrastructure. 5) Feasibility of data collection and analysis. 6) Ability to adjust for confounding factors. 7) Understanding of linkages with rest of ecosystem. 8) Broad geographic distribution. 9) Well-known life history (for fauna). 10) Nonintrusive sampling.

Legend: * = Low; ** = Low–Medium; *** = Medium; **** = Medium–High; ***** = High

Type A = for trends assessment.
Type B = for causality assessment.

TABLE 6.2
Group A: Measuring trends

Indicator	Location (intensive, cluster, or both)	Frequency
Soil solutions (THg and MeHg)	Intensive	Quarterly
Sediment (THg and MeHg)	Cluster and intensive	Annual and quarterly, respectively
Percent MeHg in sediment	Cluster and intensive	Annual and quarterly, respectively
Instantaneous methylation rate	Cluster and intensive	Annual and biannual, respectively
THg accumulation rate in sediment	Cluster	Every 5–10 years
Surface water (THg and MeHg)	Cluster and intensive	Annual and quarterly, respectively
Stream	Intensive	Weekly
THg in prey fish	Cluster and intensive	Annual
THg in mammal blood and fur[a]	Cluster and intensive	Annual
THg in bird blood, feather, and egg[a]	Cluster and intensive	Annual

[a] Species of birds and mammals vary in distribution and habitat preferences and of course can only be used as indicators if present.

6.2.2.1 Sampling Scale

This monitoring framework should be applied across broad geographic regions. This book recommends a national or (preferably) continental scale of assessment. The data collected in the United States should also be comparable, to the extent feasible and appropriate, with other North American and global mercury monitoring efforts, particularly monitoring of atmospheric transport and deposition.

To monitor a region as broad as the continental United States, it is recommended that a nested sampling approach be taken. A nested approach combines frequent, intensive monitoring for many parameters at several intensive study locations, with less frequent monitoring for additional parameters at cluster sites surrounding the intensive sites. This design is regarded as the best trade-off between collecting detailed accurate measurements at all locations and sampling in many different types of locations. To estimate regional temporal trends, at least one representative intensive site should be located in each ecoregion or type of ecosystem. The extent of human disturbance (e.g., land-use changes) in the ecosystem, as well as the atmospheric load of mercury, are also both important considerations. The cluster sites surrounding each intensive site should imbed a variety of parameters that are critical to estimating mercury transformation processes, such as dissolved organic carbon and pH for aquatic environments, and throughfall and litterfall in terrestrial environments.

This nested approach is similar to both the one used by Danz et al. (2005) to develop a stratified sampling design for general environmental indicator monitoring in the Great Lakes and the framework used in the Temporally Integrated Monitoring of Ecosystems and Long-Term Monitoring (TIME/LTM) programs. The TIME/LTM monitoring sites were carefully chosen to allow extrapolation of data on lake and stream acidification trends to surrounding areas or regions, and are resampled

TABLE 6.3
Group B: Assessing causality

Indicator	Location (intensive, cluster, or both)	Frequency
Atmospheric Hg	Intensive	Continuous
Wet deposition (THg)	Intensive	Weekly
Throughfall (THg and MeHg)	Intensive	Weekly
Litterfall (THg and MeHg)	Intensive	Weekly
Snowpack (THg)	Cluster	Annual
Soil (THg and MeHg)	Intensive	Once to characterize pools
Soils solutions (THg and MeHg)	Intensive	Quarterly
Forest floor surveys	Intensive	10 years
Sediment (THg and MeHg)	Both	Quarterly
% MeHg in sediments and soils	Both	Quarterly
Instantaneous sediment methylation rate	Intensive	Biannual
THg Accumulation in cores (THg and MeHg)	Intensive	5–10 years
Groundwater (THg and MeHg)	Intensive	Quarterly
Surface water (THg)	Both	Quarterly
MeHg/THg ratio in surface water	Both	Quarterly
Raccoon	Cluster (when feasible)	Annual
Mink	Both (when feasible)	Annual
Harbor porpoise	Cluster (when feasible)	Annual
Common loon	Both (when feasible)	Annual
Herring gull	Cluster (when feasible)	Annual
Common tern	Cluster (when feasible)	Annual
Leach's storm-petrel	Cluster (when feasible)	Annual
Belted kingfisher	Both (when feasible)	Annual
Tree swallow	Both (when feasible)	Annual
Bicknell's thrush	Cluster (when feasible)	Annual
Saltmarsh sharp-tailed sparrow	Cluster (when feasible)	Annual

Note: Species of birds and mammals vary in distribution and habitat preferences and of course can only be used as indicators if present.

regularly to provide information on regional trends (Kahl et al. 2004). The authors believe a similar monitoring framework would be effective in assessing trends in mercury contamination of the recommended indicators.

While an exact number of site clusters has not been proposed, the authors consider from approximately 3 to 10 clusters of sites to be appropriate. These clusters should represent different ecoregions with different ecological characteristics as well as different loadings (both in amount and source) of mercury deposition. Care should be taken to monitor different types of water bodies and watersheds (e.g., seepage lakes, drainage lakes, old reservoirs, rivers, and estuaries). Areas that should be considered as potential cluster site locations include lakes in northern New England/the Adirondacks, lakes in the upper Midwest, rivers and streams in the southeastern coastal plain, lakes in south-central and southeastern Canada, western mountain lakes, streams and reservoirs, the Arctic region of Alaska, and large coastal estuaries (e.g., Chesapeake Bay, Mobile Bay). The monitoring locations within each geographic region should represent the most common type(s) of ecosystem (e.g., lake, river, estuary) within that ecoregion.

Each cluster should contain a minimum of 15 to 20 monitoring sites. Several criteria should be considered when selecting individual monitoring sites within each cluster. First, the sites should represent the range of important ecosystem character-istics that determine ecological response rate to changes in mercury load (e.g., pH, DOC, presence of wetlands in the watershed, and for lakes, hydrologic type). Highly contaminated and recently disturbed sites should be avoided for the most part, because they introduce large numbers of confounding variables in assessing the data, although some authors recommend that some monitoring take place in urban sites. In addition, sampling of biota should take into account known fishing and hunting pressures. For example, sampling of fish in water bodies subject to high fishing pressure (where the largest and most contaminated fish are frequently removed from the system) may not be the most reliable way to assess changes in fish tissue mercury (Chapter 4).

The number of intensive monitoring sites should depend on the number of ecoregions selected, with at least one intensive monitoring site included in each cluster. Monitoring at intensive sites should focus on mercury concentrations in the multimedia indicators identified in Tables 6.2 and 6.3 (atmospheric deposition, soils, sediments, water, and biota), as well as ancillary measurements needed to interpret the mercury data collected. Most of these sites should be located in areas where change in loads (and therefore ecological contamination) are expected. For compar-ison, however, at least one site should be located in an area where small changes in Hg loadings are expected.

Finally, the number of monitoring sites must provide sufficient statistical power to detect expected trends. The ability to detect concentration trends over time is sensitive to the magnitude of year-to-year variation, and awareness of the different components of variance in a system of multiple sites can aid in designing classifi-cation systems for sites that increase trend detection power. The power to detect trends in a network of multiple sites with established variation among individual components is determined by the number of sites and length of sampling time. The number of sampling sites proposed in each ecoregion (15 to 20) approximates the minimum number required to detect trends of a few percent per year as long as year-to-year variation is small (Larsen et al. 2001).

6.2.2.2 Sampling Location

The selection of monitoring sites depends on a variety of factors, from the practical aspects of access and maintenance of sites to the assessment question the sampling must answer. Given that one of the purposes is to assess responses to reductions in atmospheric load, it follows logically that some sites should be located in areas where large changes in mercury loadings are expected. Other sites should be located in areas where smaller changes in mercury loadings are expected, to provide a comparison and to minimize the impact of confounding factors. In addition, different ecosystems are expected to respond at different rates based on their biogeochemistry or other factors separate from mercury load. Tables 6.2 and 6.3 contain the details of the proposed type of sampling location (intensive, cluster, or both) for each recommended indicator. The authors recommend that clusters of sampling sites be distributed across the United States and the Canadian provinces and represent the major ecoregions currently known to be affected by mercury contamination.

6.2.2.3 Sampling Frequency

The recommended sampling frequencies in Tables 6.2 and 6.3 are based on several factors. The first, and most important, factor to consider is the question being addressed. Some questions require very frequent sampling, whereas others require much less frequent sampling.

The second factor is the temporal variation in concentrations in different ecosystem compartments. For example, sediments and prey fish exhibit less temporal variation in mercury concentration than do air or water, and thus statistically valid estimates of their status can be collected with less frequent monitoring (e.g., annual sampling for prey fish vs. daily or hourly sampling for atmospheric concentrations of mercury).

The third factor is how quickly changes are expected to occur in the monitored ecosystem compartment. For example, the concentration of mercury in water varies substantially through the year, whereas concentrations in large fish or piscivorous wildlife, such as otters, change much more slowly. The assessment program should conduct sampling often enough to capture most of the important changes taking place in the ecosystem compartments (indicators) of greatest concern.

The fourth factor affecting sampling frequency is efficiency, that is, to collect the minimum number of samples needed to address the key questions. The costs of collecting and analyzing samples are substantial, and the "oversampling" of a particular indicator would reduce resources available for monitoring other indicators or other sites.

6.2.2.4 Overall Duration of Sampling

How long should monitoring programs continue? For the purposes of specifically answering the question "What is the ecosystem response to changes in atmospheric loadings of mercury?", some specific timeframes should be considered.

Reductions in U.S. mercury emissions from medical and municipal waste incinerators and other industrial sectors have already occurred. Additional emission reductions from some coal-fired power plants have also already begun as co-benefits from technologies used to control SO_2 and NO_x emissions. These mercury emissions from power plants are, however, expected to be reduced further over the next few decades. Meanwhile, changes in mercury emissions in other parts of the world may also affect some U.S. ecosystems.

There is uncertainty concerning the length of time it will take for ecosystems to respond to reductions in mercury deposition. It is agreed, however, that it is important to continue monitoring until the majority of the ecosystem responses have been documented. If monitoring is discontinued before that point, the full response and benefit of the mercury emission reductions will not be known. The authors believe this monitoring should occur for at least 15 to 20 years to capture most of the changes expected from existing and expected U.S. reductions in atmospheric deposition of mercury. It is expected that this estimated timeframe will be refined as the monitoring progresses and the response of ecosystems is assessed.

6.2.3 MONITORING FOR TRENDS AND MONITORING FOR CAUSALITY

The indicators recommended in the previous chapters and identified in Table 6.1 have been assigned to 1 of 2 categories: A indicators (to measure trends) and B indicators (to examine the role of changes in atmospheric mercury emissions and deposition). The purpose of this categorization is to provide guidance to policymakers and other stakeholders regarding the most important environmental mercury monitoring needs. It is also designed to provide guidance regarding which types of indicators are most useful for answering specific policy questions.

Given that atmospheric deposition is only one of several factors that control the amount of methylmercury accumulated in fish, a monitoring program focusing only on indicators such as Hg deposition and fish mercury levels cannot explain the differences that will undoubtedly arise when the data sets are compared. It also cannot explain ecosystem contamination in higher trophic levels (e.g., piscivorous birds and mammals). To fully understand the extent to which reductions in atmospheric deposition are responsible for reductions in fish tissue concentrations of mercury, process-based studies and modeling are needed, as well as additional monitoring information. These additional data are identified as B indicators. The ancillary data (e.g., precipitation, pH) that must be collected to interpret both types of indicators are identified in Table 6.4.

6.2.4 INTEGRATION OF MONITORING WITH MODELING CAPABILITIES

This book proposes a monitoring program that will help determine trends for mercury concentrations in the environment and assess the relationship between these concentrations and mercury emissions. Environmental models are also often used to predict trends and examine relationships among variables. Models can facilitate the interpretation of data emerging from monitoring programs recommended in this book and that the data will help develop better modeling tools.

Our understanding of mercury cycling and bioaccumulation has advanced markedly in recent years, but gaps still remain (USEPA 2002). The same can be said for the application of that information to mercury simulation models. Resolving some key issues, such as those regarding atmospheric cycling, terrestrial export of mercury, and factors governing methylation and bioaccumulation, would result in stronger models (Atkeson et al. 2003). The scientific understanding and models of mercury behavior in terrestrial catchments (including wetlands) are currently both less developed than their aquatic and atmospheric counterparts.

Results from the monitoring approach recommended by this book, particularly from intensively studied sites, offer the potential to further advance the development of models of mercury cycling. The data from these sites can provide critical information

TABLE 6.4
Ancillary monitoring parameters

Parameter	Location (intensive, cluster, or both)	Frequency	Type (A or B)[a]
Atmospheric deposition of sulfate	Intensive and cluster	Weekly	B
Rainfall	Intensive and cluster	Weekly	B
Watershed area	Intensive and cluster	Once	B
Land cover	Intensive and cluster	Once	B
Percent wetlands in watershed area	Intensive and cluster	Once	B
Lake morphometry	Intensive and cluster	Once	B
Water chemistry (pH, DOC, sulfate, TSS, chlorophyll, temperature, ANC, color, nutrients, DO, stratification status)	Intensive and cluster	Quarterly	A, B
Characteristics of fish (size, age, stomach contents, sex, condition)	Intensive and cluster	Annual for prey fish; 1- to 3-year intervals for piscivorous fish	A, B
Characteristics of mammals (size, age, sex, condition) and tissues for nonlethal sampling (fur and blood) and lethal sampling (fur, brain, muscle, liver)[b]	Intensive and cluster	Annual	A, B
Characteristics of birds (size, age, sex, condition) and tissues for nonlethal sampling (blood, feathers and eggs) and lethal sampling (feathers, brain, muscle, liver and eggs)[b]	Intensive and cluster	Annual	A, B

[a] Type A = for trends assessment; Type B = for causality assessment.
[b] Species of birds and mammals vary in distribution and habitat preferences, which will determine their suitability at particular sites.

to test hypotheses and improve the models. Data from the larger number of sites involving limited sampling would probably be less well suited to model development, but could in some cases be used to test the predictive strength of models that have been already developed and calibrated.

It would also be prudent to consider opportunities to use models to interpret results emerging from intensively studied sites. Models can provide frameworks to integrate and examine complex data sets in multidisciplinary studies, and can identify key information needed in field programs. Models have been successfully used in this manner in several comprehensive mercury research programs — for example, the Mercury in Temperate Lakes program in the early to mid-1990s in Wisconsin (Hudson et al. 1994) and the Aquatic Cycling of Mercury in the Everglades (ACME) research program in the Florida Everglades from the mid-1990s to the present (Atkeson et al. 2003). Models may also help isolate the effects of a single factor, such as mercury deposition, among the many factors that can affect the fate and transport of mercury.

6.3   COMPLEXITIES/CONFOUNDING FACTORS

As noted many times in this book thus far, there are many confounding factors related to the complexity of ecosystems that complicate the interpretation of monitoring data. For example, inter-annual weather and population variability (e.g., droughts, cold winters, plankton blooms) can significantly affect ecosystem processes and mercury cycling. Anthropogenic disturbances, such as deforestation, urbanization, and the creation of reservoirs, can all significantly complicate our ability to identify trends in mercury contamination. Human activities that directly affect biota, such as fishing and the introduction of exotic species, can also complicate the assessment process. Finally, changes in mercury loads are taking place against a background that includes changing rates of regional sulfate deposition and global climate change.

The monitoring framework should be designed to minimize the influence of such factors or to control for them in the analysis process. While no monitoring strategy can account for all possible complexities, the authors believe this monitoring framework has identified the best currently available solutions to address problems created by these confounding factors.

6.4   RECOMMENDATIONS

This book has identified the most useful indicators of environmental changes in mercury contamination in 4 compartments of the environment: 1) airsheds and watersheds, 2) water and sediment, 3) aquatic organisms (with emphasis on fresh-water ecosystems), and 4) wildlife that live in freshwater, terrestrial, and/or coastal ecosystems. The indicators identified in this book are wide-ranging and involve measurements made at several different scales of time and space. The authors believe that these indicators will provide the best information to policymakers, as well as other stakeholders, as to whether environmental concentrations are changing (A indicators) and what the reasons for those changes might be (B indicators).

To assess changes in mercury concentrations in the environment, it is important to have a baseline from which to measure change. Due to the fact that mercury emissions have already been reduced from a number of sectors in the United States, and that further emission reductions are expected from power plants over the next few decades, it is critical that an assessment program be implemented soon.

Therefore, it is strongly recommended that federal and state agencies and other interested partners begin the process of designing and implementing an ecological mercury monitoring program as soon as possible.

To adequately assess ecological changes in mercury concentrations, it will be necessary to make a long-term commitment to ecological monitoring. The assessment questions will not be answered in 3 years or even in 5 years; rather, it will likely take at least 15 to 20 years before the full scope of the impacts of the emission reductions are measured, depending on the types of systems being monitored. For this reason, it is critical for federal agencies and other cooperators to make a firm commitment to support this monitoring effort for many years into the future. The authors recognize the fiscal realities that limit long-term planning but do believe that it is possible for federal agencies, states, private-sector organizations, and others to provide long-term support for the proposed monitoring program.

It is recommended that all organizations participating in the ecological mercury assessment program make formal commitments, subject to budget constraints, to recognize and support the long-term continuation of this monitoring program.

The monitoring program described in this book is a substantial undertaking that will require a sustained commitment and substantial resources. The authors view this monitoring program as too large and complex to be housed in any one particular federal agency or academic institution. Rather, it will require participation from many federal and state agencies, as well as Indian tribes and universities. A good model for the monitoring program may be the National Atmospheric Deposition Program (NADP), a consortium of more than 250 sponsors that depends on substantial funding support from at least a half-dozen different federal agencies and departments as well as guidance from a management committee composed of scientists from a wide range of backgrounds and organizations

It is recommended that federal and state agencies work with each other and with other interested organizations to develop a broad consortium to support and guide this monitoring and assessment program.

REFERENCES

Atkeson T, Axelrad D, Pollman C, Keeler G. 2003. Integrated atmospheric mercury deposition and aquatic cycling in the Florida Everglades: an approach for conducting a total maximum daily load analysis for an atmospherically derived pollutant. Florida Department of Environmental Protection. 95 p.

Danz NP, Regal RR, Niemi GJ, Brady VJ, Hollenhorst T, Johnson LB, Host GE, Hanowski JM, Johnston CA, Brown T, Kingston J, Kelly JR. 2005. Environmentally stratified sampling design for the development of Great Lakes environmental indicators. Environ Monitor Assess 102:41–65.

Engstrom DR, Swain EB. 1997. Recent declines in atmospheric mercury deposition in the upper Midwest. Environ Sci Technol 31:950–967.

Fevold BM, Meyer MW, Rasmussen PW, Temple SA. 2002. Bioaccumulation patterns and temporal trends of mercury exposure in Wisconsin common loons. Ecotox 12:83–93.

Fitzgerald WF, Engstrom DR, Mason RP, Nater EA. 1998. The case for atmospheric mercury contamination in remote areas. Environ Sci Technol 32:1–7.

Griffith LM, Ward RC, McBride GB, Loftis JC. 2001. Data analysis considerations in producing 'comparable' information for water quality management purposes. Available at: http://water.usgs.gov/wicp/acwi/monitoring/pubs/tr/nwqmc0101.pdf

Grimm JW, Lynch JA. 1991. Statistical analysis of errors in estimating wet deposition using five surface estimation algorithms. Atmos Environ 25(2):317–327.

Hrabik TR, Watras CJ. 2001. Recent declines in mercury concentration in a freshwater fishery: isolating the effects of de-acidification and decreased atmospheric mercury deposition in Little Rock Lake. Sci Tot Environ 297:229–237.

Hudson RJM, Gherini SA, Watras CJ, Porcella DB. 1994. Modeling the biogeochemical cycle of mercury in lakes: the mercury cycling model (MCM) and its application to the MTL study lakes. In: Watras CJ, Huckabee JW, editors, Mercury pollution: integration and synthesis. Boca Raton (FL): Lewis Publishers, p. 473–523.

Jackson TA. 1997. Long-range atmospheric transport of mercury to ecosystems, and the importance of anthropogenic emissions — a critical review and evaluation of the published evidence. Environ Rev 5:99–120.

Kahl JS, Stoddard JL, Haeuber R, Paulsen SG, Birnbaum R, Deviney FA, Webb JR, Dewalle DR, Sharpe W, Driscoll CT, Herlihy AT, Kellogg JH, Murdoch PS, Roy K, Webster KE, Urquhart NS. 2004. Have U.S. surface waters responded to the 1990 Clean Air Act Amendments? Environ Sci Technol 38(24):484A–490A.

Kamman NC, Engstrom DR. 2002. Historical and present fluxes of mercury to Vermont and New Hampshire lakes inferred from ^{210}Pb dated sediment cores. Atmos Environ 36:1599–1609.

Kincaid TM, Larsen DP, Urquhart NS. 2004. The structure of variation and its influence on the estimation of status: indicators of condition of lakes in the Northeast, U.S.A. Environ Mon Assess 98:1–21.

Larsen DP, Kincaid TM, Jacobs SE, Urquhart NS. 2001. Designs for evaluating local and regional scale trends. Bioscience 51:1069–1078.

Olsen AR, Sedransk J, Edwards D, Gotway CA, Liggett W, Rathbun S, Reckhow KH, Young LJ. 1999. Statistical issues for monitoring ecological and natural resources in the United States. Environ Mon Assess 54:1–45.

Schindler DW, Kidd KA, Muir DCG, Lockhart WL. 1995. The effects of ecosystem characteristics on contaminant distribution in northern freshwater lakes. Sci Tot Environ 160/161:1–17.

Stow CA, Carpenter SR, Webster KE, Frost TM. 1998. Long-term environmental monitoring: some perspectives from lakes. Ecol App 8:269–276.

[UNEP] United Nations Environment Programme. 2002. Global mercury assessment. Geneva, Switzerland: UNEP Chemicals. 270 p.

[USEPA] US Environmental Protection Agency. 1997. Mercury study report to Congress. Washington, D.C.: USEPA, Office of Air Quality Planning and Standards and Office of Research and Development. EPA 452/R-97-003.

[USEPA] US Environmental Protection Agency. 2002. Proceedings and summary report, workshop on the fate, transport, and transformation of mercury in aquatic and terrestrial environments. EPA/USGS workshop; 2001 May 8–10; West Palm Beach, FL, USA. USEPA Office of Research and Development. EPA/625/R-02/005. 171 p.

Urquhart NS, Paulsen SG, Larsen DP. 1998. Monitoring for policy-relevant regional trends over time. Ecol App 8:246–257.

Wiener JG, Krabbenhoft DP, Heinz GH, Scheuhammer AM. 2003. Chapter 16: Ecotoxicology of methylmercury. In: Hoffman DJ, Rattner BA, Burton Jr GA, Cairns Jr J, editors, Handbook of ecotoxicology, 2nd ed. Boca Raton (FL): CRC Press, p. 407–461.

Index

A

ACME project, 63, 73, 204
Air pollution, 26, 39
Air quality
    ambient, 18, 20, 21
    Hg intensive sites, 32–33
Airsheds, 13–14; *see also* Atmospheric mercury
    concentrations of Hg species, 24
    defining, 22
    response to changes in Hg emissions, 14
    sampling, 25–26
    spatial variations, 35,37
Alaska Marine Mammal Tissue Archival Project
    (AMMTAP), 159
Albatrosses, 146
Algae, 98–100
Alligators, 142
American dipper, 142
Amphibians, 143, 145
Ancillary data, *see* Data
Anthropogenic emissions, *see* Emissions
Aquatic biological indicators, 88, 91–92; *see also*
    Fish; Trophic position
    ancillary data, 107
    confounding factors, 88, 90, 94, 108–110
    enhancing interpretations, 110–113
    fish, 92–95
    intrinsic co-variables, 90
    periphyton, 99–100
    phytoplankton, 98–99
    recommended, 100, 101–103
    selection criteria, 90–91
    trend monitoring, 104–107, 111–112
    zooplankton, 97–98
Aquatic biota, *see* Aquatic biological indicators
Aquatic Cycling of Mercury in the Everglades
    (ACME), 63, 73, 204
Aquatic systems; *see also* Aquatic biological
    indicators; Reservoirs; Rivers; Surface
    waters; Watersheds
    controlling factors of Hg concentrations, 70, 77
    and dissolved organic matter (DOM), 71–72
    disturbances to, 81
    estuaries, 88, 96, 97
    exotic species, 109
    historical data, 106–107

MeHg in water, 75–78
methylation, 108
point-source discharges, 105, 106
recommended water indicators, 52–54
response to altered mercury loading, 47–50
temporal trends of HgT, 72, 73
trophic levels, 108–109
water chemistry, 80–81, 199
Atmospheric deposition, 130; *see also*
    Atmospheric mercury; Temporal response
    climatic factors, 16, 17, 18
    estimation of total deposition, 33, 34, 35
    in Florida Everglades, 163
    geographical patterns, 29–31
    at high latitudes, 130
    measurement sites, 32–33
Atmospheric mercury, 17, 196; *see also*
    Atmospheric deposition
    existing monitoring networks, 27–31
    modeling, 27
    residence times, 22, 26
    sampling, 26–27
    sources, 15
    species, 22–23, 24

B

Background levels of Hg, 22, 30
Bald eagles, 131, 137
Baselines, 9, 192
Bass, 73, 104, 162
Bats, 135–136, 144, 145
Belted kingfishers, 124, 128, 140–141, 145, 146
Benthic invertebrates, 88, 91, 95–97, 100
trophic status, 106
Bicknell's thrush, 141
Bioaccessibility, 74, 77, 125–126
Bioaccumulation, 2, 87, 89, 91–92, 203; *see also*
    Aquatic biological indicators
    adult survivorship endpoints, 134
    in benthic invertebrates, 96
    factors in water quality criteria model, 147
    and mercury burden, 106
    models of, 9
    and sediment-based indicators, 55
    watershed influences, 112